Pieces of Mind

Pieces of Mind

The Proper Domain of Psychological Predicates

Carrie Figdor

OXFORD
UNIVERSITY PRESS

OXFORD
UNIVERSITY PRESS

Great Clarendon Street, Oxford, OX2 6DP,
United Kingdom

Oxford University Press is a department of the University of Oxford.
It furthers the University's objective of excellence in research, scholarship,
and education by publishing worldwide. Oxford is a registered trade mark of
Oxford University Press in the UK and in certain other countries

First Edition published in 2018

Impression: 1

Published in the United States of America by Oxford University Press
198 Madison Avenue, New York, NY 10016, United States of America

British Library Cataloguing in Publication Data

Data available

Library of Congress Control Number: 2017958148

ISBN 978-0-19-880952-4

Printed and bound by
CPI Group (UK) Ltd, Croydon, CR0 4YY

In memory of
Walter Figdor, Rae Figdor, and Guacaipuro Montesinos

Contents

Acknowledgments

This book began as an attempt to understand mechanistic explanation of mind as it was being implemented in the cognitive sciences. The bulk of it consists of examining the first pieces of that puzzle—the ones that usually escape notice, as Wittgenstein might say. The result is an example of what can happen when you approach an issue without aiming at a particular conclusion. It is also an example of what happens when other people want to know what you're doing while you're still figuring things out. Initial exploratory feints presented at the Society for Philosophy and Psychology annual meeting in summer 2009 (at Indiana University–Bloomington) and at Ruhr-Universität Bochum in summer 2013 (courtesy of "What is Cognition?" workshop organizers Cameron Buckner, Ellen Fridland, and Albert Newen) have little superficial resemblance to the penultimate draft discussed at the University of Edinburgh in summer 2016 in a workshop organized by Mark Sprevak (whose support has been particularly noteworthy). But it really was the same project all along, even if a large amount of research in empirical and philosophical literatures didn't make it into the book.

Academic year 2013–14, as a fellow at the University of Pittsburgh Center for Philosophy of Science (then directed by John Norton), was invaluable for providing me with the time, support, and informed critical pushback I needed at just the right time. I was greatly helped by interactions with fellows Aristides Arageorgis, Ori Belkind, Melinda Fagan, Allan Franklin, Marco Giovanelli, Leah Henderson, Arnaud Pocheville, and Joshua Rosaler, as well as Pittsburgh faculty and students Jim Bogen, Trey Boone, Mazviita Chirimuuta, Peter Machamer, Joe McCaffrey, and Mark Wilson. I also thank the University of Iowa for financial support for my year at Pittsburgh and for travel grants to present parts of the project at other institutions.

In addition to my colleagues at the University of Iowa—including apt historical references suggested by Ali Hasan and Carrie Swanson, and students Mark Bowman, Thomas Butler, Hyungrae Noh, and Derek Voel-Pel—I have had helpful responses (and much pushback) on various issues raised within this book from audiences at the CUNY Graduate

Center Cognitive Science program, Durham University, Hebrew University Institute for Advanced Study, ENS-Institut Nicod, Northwestern University, Stanford University, University of California–Riverside, University of Colorado–Boulder, University of Manchester, University of Melbourne, Washington University at St. Louis, and Indiana University. I also thank audiences at presentations of some of the material at meetings of the American Philosophical Association, Aristotelian Society/Mind Association, Australasian Association for Philosophy, Society for Philosophy and Psychology, and Southern Society for Philosophy and Psychology.

In addition, many individuals offered insightful comments and raised hard questions every step of the way. These include two anonymous reviewers at Oxford University Press, whose questions helped me strengthen the book throughout. Others who commented on draft chapters or aspects of the project include Mikio Akagi, Colin Allen, Michael Anderson, Brice Bantegnie, Bill Bechtel, Ian Bisset, David Chalmers, Andy Clark, Carl Craver, Bryce Huebner, Tom Kane, Fred Keijzer, Uriah Kriegel, Suilin Lavelle, Jonny Lee, Keya Maitra, Corey Maley, Ron Mallon, Christopher Mole, Carlos Montemayor, Gualtiero Piccinini, Joelle Proust, Miza Rashid, Robert Rupert, Eric Schwitzgebel, Georg Theiner, Michael Weisberg, and Wayne Wu. I would also like to thank Peter Momtchiloff at OUP for his help guiding me through the publication process; the OUP editorial staff to whom this book was assigned; and my indexer, Lacey Davidson. If I've left someone out, I regret the omission and hope to make up for it in the future.

Finally, I would like to thank Bill, Liz, and Tom for their indispensable background support.

1

Introduction

1.1 General Remarks

Great cases, like hard cases, make bad law. For great cases are called great, not by reason of their real importance in shaping the law of the future, but because of some accident of immediate overwhelming interest which appeals to the feelings and distorts the judgment.

(*N. Sec. Co. v. United States*, 193 U.S. 197, 400 (1904)
(Holmes, J., dissenting))

This book is an attempt to understand a specific and ongoing process of conceptual revision: the revision in psychological concepts in response to new discoveries throughout the life sciences. Because the claim that a conceptual transition is going on is itself contested, the book shows why it is plausible as well as how uses of psychological predicates should plausibly be interpreted in its light. I call the view I hold Literalism. When scientists report that fruit flies and plants decide, bacteria communicate linguistically, and neurons prefer, the Literalist claims that the most plausible interpretation of these unexpected uses of psychological predicates is the Literal one, explained more fully in what follows. This interpretation reflects the empirically driven revision in our understanding of the nature of the properties to which they refer. My goals in this book are to defend Literalism and explore its implications for the philosophical projects of naturalizing the mind and understanding the role of the mind in grounding moral status.[1]

---

[1] In Figdor (2017) I labeled my view Anti-Exceptionalism without clearly distinguishing its metaphysical and semantic components. I distinguish them here: Literalism is the overall view, and Anti-Exceptionalism is its metaphysical component. Also, in this introduction I use the terms "predicate", "term", and "concept" interchangeably. In subsequent chapters, I will draw distinctions between them where necessary.

Although conceptual change is a venerated topic within philosophy of science, I have three main motivations for examining this particular transition for a broader audience. First, the nature of mental capacities is intrinsically interesting to almost everyone—unlike, say, the nature of acidity (Chang 2012). The topic of psychological conceptual change inherits this interest. Second, there is a pressing need to make sense of scientists' extensive, systematic, and selective use of psychological language to explain new discoveries of biological complexity.[2] Some philosophers have made important contributions to this project by discussing specific cases of the issue (e.g. Bennett and Hacker 2003 and Bennett et al. 2007 in cognitive neuroscience, Calvo and Keijzer 2009 in plant science). What is needed is recognition of the pervasiveness of the phenomenon and, if possible, a unified account of it that shows why researchers across a wide range of biological sciences are rationally justified in using the terms to describe phenomena in their fields. Third, cognitive researchers are actively seeking to reveal the patterns of brain activity supporting mental function at numerous levels of organization and complexity and to develop the conceptual framework adequate for characterizing these patterns (e.g. Price and Friston 2005; Carandini 2012; Poldrack and Yarkoni 2016). This research may be mistakenly constrained if we fail to question the received view of the conditions on an adequate scientific explanation of the mind. As articulated in philosophy of mind and psychology, this view claims that even if psychological terms are not eliminated from scientific discourse (Churchland 1981), a naturalistic explanation requires that they be translatable into or replaceable in serious scientific contexts by a non-psychological vocabulary. Paraphrasing Fodor (1987: 97), if psychological properties are real, they must really be something else. But there may be a better way to understand what a naturalistic explanation of mind requires.

From the perspective defended here, what is occurring now with psychological concepts (and terms) is what has already happened with many others: we are accommodating the concepts and the theories in which they figure to new empirical knowledge. This transition is not

---

[2] I focus here on biological entities, setting aside artificial intelligent agents and other cases for future work (though I do briefly consider them where relevant). The assumed biological background is strategic; I do not use it to draw a conceptual distinction.

instantaneous. Just as it took time to accept geometry as providing a literal account of reality (Pylyshyn 1993: 555–6), it will also take time to adjust, and adjust to, the actual extensions of psychological predicates. What makes this case more contentious than other scientifically-driven revisions of conceptual frameworks is that it entails that the standards for proper use are being detached from their traditional home in specifically human behavior and cognition. The end state of this transition is the end of psychological and psychological-conceptual anthropocentrism. The quotation above from U.S. Supreme Court Justice Oliver Wendell Holmes expresses this perspective. Human minds are great cases, but they are not the standard for determining the proper extensions of psychological predicates. As with much of nature, what is most accessible and familiar to us does not ground an accurate account of the phenomena.[3] How this separation is occurring—even how it *can possibly* occur—will be clarified in the rest of this book.

For many readers, it may be helpful to relate this transition to Wilfrid Sellars' (1963/1991) well-known discussion of scientific progress and its relation to the human self-conception. In *Philosophy and the Scientific Image of Man*, Sellars presents two idealized, and at least apparently competing, conceptions of man-in-the-world, dubbed the Manifest and Scientific Images. The Manifest Image is a conceptual scheme of people and things in which human beings are exceptional in the world of things. The Scientific Image is a conceptual scheme in which we introduce theoretical terms, referring to theoretical entities, to explain the things and behaviors present to us in the Manifest Image. So much is broadly familiar.

Less familiar is the fact that the Manifest Image itself was a "refinement" or "modification" of an earlier conceptual scheme, the Original Image. This image is "a framework in which all the 'objects' are persons":

an image in which all the objects are capable of the full range of personal activity, the modification consisting of a gradual pruning of the implications of saying with respect to what we would call an inanimate object, that it did something.

(Sellars 1963/1991: 12)

---

[3] The unadulterated lines from Aristotle are from *Physics* Book I: "The natural course is to proceed from what is clearer and more knowable to us, to what is more knowable and clear by nature; for the two are not the same. Hence we must start thus with things which are less clear by nature, but clearer to us, and move on to things which are by nature clearer and more knowable."

Sellars adds:

From this point of view, the refinement of the 'original' image into the manifest image is the gradual 'de-personalization' of objects other than persons. That something like this has occurred with the advance of civilization is a familiar fact. Even persons, it is said (mistakenly, I believe), are being 'depersonalized' by the advance of the scientific point of view. (1963/1991: 10)

Given Sellars' characterization of the transition from the Original to the Manifest Image, it would appear that pruning the person-appropriate predicates and pruning the psychological predicates are one and the same exercise.[4] However, scientific discoveries are showing that the intuitive chasm between the behavioral flexibility and intelligence of human beings and the behavioral inflexibility and stupidity of nonhuman beings is empirically unsustainable. This chasm presumably helped motivate the pruning process; it certainly helps to maintain the Manifest Image. But just as the rampant anthropomorphism of the Original Image gave way to the Manifest Image, the latter's psychological anthropocentrism is giving way to a Scientific Image in which the psychological and the human (the "person-appropriate") are decoupled and conceptually separable, yet not thereby dehumanized or dehumanizing.

Literalism is one response to these developments. On Sellars' view, traces of the Original Image linger in the Manifest Image in uses of such predicates in relation to nonhumans "for poetic and expressive purposes", such as when ascribing processes of deliberation to nonhumans and describing their doings as actions. For example, "the use of the term 'habit' in speaking of an earthworm as acquiring the habit of turning to the right in a T-maze, is a metaphorical extension of the term" (Sellars 1963/1991: 12). On the Literalist view, biologists have not become alarmingly poetic in their peer-reviewed publications or professional presentations over the past few decades. When they say that worms can acquire habits, they mean that worms can acquire habits. In fact, they can be classically conditioned: what scientists now are most interested in finding out about *C. elegans* is not so much what species members can learn, but what they

---

[4] For present purposes, 'persons' (of the non-legal sort) and 'humans' may be considered co-extensive. The metaphysical category of personhood (as contrasted with a social or evaluative category) is associated with those with human bodies (Sapontzis 1981), while contemporary researchers in dehumanization often employ the terms 'depersonalization' and 'dehumanization' synonymously (e.g. Loughnan et al. 2010).

can't (Rankin 2004). Scientific progress is yielding the opposite of what eliminativism in philosophy of mind (Churchland 1981) predicts or recommends. From a Literalist perspective, it is a contingent fact that it has taken so long for us to be in a position to test scientifically whether the drastic pruning characteristic of the transition from the Original to the Manifest Image is based on a fundamental mistake.

As we shall see, metaphorical or otherwise non-Literalist interpretations of these uses remain live options. But these interpretations must be hard-won. For the Literalist position rests on the familiar and widely accepted idea that science has priority over commonsense intuition in determining the extensions of predicates denoting natural phenomena. Science is eliminating the anthropocentrism from the psychological, not eliminating the psychological. Intuitions based on subjectively familiar phenomena remain an essential source of evidence about how human capacities appear to us and an important, if parochial, source of evidence about their nature. Nevertheless, a shift away from psychological anthropocentrism implies that the conceptual adjustments will be assimilated into the familiar human domain as long as we continue to use these concepts to characterize ourselves and explain our behavior. This is likely, given their utility.

The conceptual change I have in mind involves a revision or precisification of reference to reflect new scientific discoveries regarding psychological capacities or properties. The change is analogous to the way scientific discovery and theorizing enabled us to determine a precise characterization of the nature of gold and, in consequence, to precisify our referential intentions and modify our uses of "gold". Such scientifically driven conceptual changes have metaphysical as well as semantic components. To a first approximation, my metaphysical position—Anti-Exceptionalism—holds that all the relevant scientific evidence shows that psychological capacities are possessed by a far wider range of kinds of entities than often assumed. We are still finding out which capacities and which entities, but the trend is unmistakable—even when focusing on biological entities, as I do in this book. (This important restriction will be discussed further in Chapter 4.) Literalism claims that, in contexts standardly interpreted as fact-stating, uses of psychological predicates to ascribe capacities to entities in this wider range are best interpreted as literal with sameness of reference. Anti-Exceptionalism is the metaphysical position that underwrites the claim of sameness of reference. I am a Literalist about the best interpretation of psychological predicates used in these contexts because the best overall understanding of the relevant scientific evidence points to Anti-Exceptionalism.

I call the overall view Literalism because it most clearly indicates what we mean when we interpret scientific research claims as literal. Of course, for non-psychological terms, this interpretation is the default; we don't point it out, or are not even aware of it, unless the context of use within a scientific paper indicates a non-literal interpretation. In contrast, uses of psychological terms for many unexpected non-human cases are often automatically interpreted as non-literal. For many, a nonhuman context of use—say, a sentence ascribing habits to earthworms—itself indicates a non-literal interpretation. When examples of the new uses are pointed out, the immediate response is that the terms are not meant literally. But this is so only if some form of Exceptionalist metaphysics for psychology is presupposed. This is the presupposition being undermined by biological discovery. Moreover, because an Exceptionalist metaphysics is often presupposed, semantic concerns often crowd out metaphysical ones: if the language is just metaphorical, why raise metaphysical issues, let alone consider the possibility of conceptual change? From my opponents' perspective, a semantic answer settles the matter; from mine, it motivates debate about the nature of psychological properties.

1.2 Chapter Summaries

The book is organized as follows. I begin by discussing the evidence supporting an Anti-Exceptionalist metaphysics, turn next to Literalism and a series of specific semantic alternatives to it, and end with consideration of some implications of Literalism. In Chapters 2 and 3, I present some unexpected uses in some detail. These are divided into extensions based on qualitative or else quantitative analogy. The former class helps illustrate the proliferation of uses across unexpected domains, and makes vivid the need to reconsider what would be an adequate account of the reference of psychological predicates. The latter class adds extensions of predicates by means of the use of mathematical models across intuitively disparate domains. The modeling practices provide a strong, new motivation to deny the anthropocentric metaphysics behind the received semantics of psychological predicates.

In Chapter 4, I present an initial defense of Literalism. I argue that it is the only position that takes seriously the dominant investigative strategies in science and that offers scientists a viable theoretical position supporting the rationality of their uses in contexts in which their default intention is to report truths.[5] I also present and reply to a series of intuition-based objections to a Literal construal of the uses, and argue that none of them succeeds in showing that Literalism is implausible.

In Chapters 5 through 7, I consider the main semantic rivals to Literalism and show why each fails to provide a satisfactory account. The alternatives are more clearly distinguished here than they usually are in the literature or have been in conversation. In many cases they have not been articulated or defended at length at all. This is probably because the basic problem has not been widely recognized. What attention there is tends to be limited to specific concepts in specific cases. These chapters also provide further opportunities to elaborate Literalism by contrasting it with these positions.

The Nonsense view (Chapter 5) has been championed most prominently by Maxwell Bennett and P. M. S. Hacker (2003; Bennett et al. 2007). Although their discussion is limited to certain kinds of uses in cognitive neuroscience, it is the only non-Literalist position that has been defended at length in print. It is not, however, the strongest alternative. The Nonsense view fails to capture the pervasiveness of the new uses and to register the full range and kinds of evidence we have for the new ascriptions.

The Metaphor view (Chapter 6) is mooted by a number of authors, including Sellars (1963/1991) and Searle (in Bennett et al. 2007). For many, it is a reflexive go-to position rather than the result of careful consideration and application of theories of metaphor to these cases. I show that this view has no independent evidence in its favor, and that our current best theories of metaphorical interpretation—the classical Gricean view, and the Relevance-Theoretic (or pragmatic semantics)

---

[5] This does not mean their default intention is to be *realists* about scientific discourse. Strictly speaking, Literalism holds that the uses of psychological terms are univocal across human and nonhuman contexts, however one accounts for the truth conditions (or success conditions, if one thinks science does not aim at truth) of sentences containing them. However, I do assume realism throughout, leaving to an anti-realist Literalist the task of framing the debate in terms appropriate to her version of anti-realism.

view—do not support the idea that the uses are intended metaphorically. I also distinguish semantic metaphor from epistemic metaphor, or the use of analogy in scientific theory building. Epistemic metaphor is consistent with Literalism.

Technical views (Chapter 7) hold that the terms are used literally but refer to distinct phenomena when used for nonhumans (and hence are not Literal). This position subdivides into two main variants. The Technical-Behaviorist variant claims that the uses refer merely to behavior or stimulus-response patterns. The Exsanguinated Property variant claims that the terms refer to "hemi-demi-semi-proto-quasi-pseudo" cognitive properties (to use Dennett's phrase, in Bennett et al. 2007: 88–9) or otherwise less-than-stellar properties. I argue that both variants fail to adequately account for the new uses. The Technical-Behaviorist variant fails to justify its referential distinction between humans and nonhumans, while the Exsanguinated Property variant introduces rather than dispels mystery, and for no good reason.

In Chapters 8 and 9, I consider two implications of Literalism that might be thought to make it undesirable even if it is the best interpretation of psychological predicates. In Chapter 8, I discuss its implications for naturalizing the mind. With respect to the dominant strategy of mechanistic or homuncular functionalist explanation of the mind, I argue that the so-called homuncular fallacy is not a fallacy. It is not a fallacy in general for mechanistic explanation outside of psychology, and attempts to justify an exception for psychological explanation fail. I also offer a new way to understanding the demand that a naturalistic explanation of mind requires psychological concepts to be discharged.

In Chapter 9, I discuss the implications of Literalism for our fears that it undermines our moral status, given the traditional role of psychological properties in grounding moral status. Literalism entails that we have more psychological features in common with more living entities than we have thought and perhaps like to think. My argument highlights the role of psychological ascriptions in drawing and maintaining social and moral boundaries, as shown most clearly in dehumanization. Even if we are not uniquely unique (Knowles 2002: 202), Literalism does not entail that we cannot continue to draw distinctions in moral status between ourselves and other entities, or that we are less deserving of high moral status. For it does not touch our motivation for using psychological ascriptions for drawing social boundaries, and there are ways for us to avoid radical adjustment of current moral-status distinctions. However, these distinctions may become increasingly arbitrary if we try to maintain them in the face of new scientific discoveries. As a result, in the longer run Literalism motivates us to find non-anthropocentric grounds for distinctions in moral status.

Chapter 10 is a concluding summary of the book. I also point out the benefits of Literalism for promoting theoretical inquiry and scientific investigation, contributing to ongoing debates about animal cognition, artificial intelligence, and group cognition, and to reframing the way we explain the behavior of ourselves and nonhumans.

1.3 Consciousness and Content

At various points in the book, I note how alternative views often beg the question by presupposing that intuitions about a special class of adult human cases are metaphysically (referentially) transparent. There is significant resistance to the serious challenge science is presenting to anthropocentric psychology. So it is worth emphasizing that even if Literalism is true, phenomenology—how mindedness feels and appears to us—has an ineliminable role to play in understanding human minds and guiding scientific investigation. Core philosophical debates about mind, language, and their relation can often proceed unchanged once a domain restriction to humans is made explicit. As a result, I will not be discussing at length two issues that loom large in philosophy of mind and language: consciousness and content. So I'll say a bit here about each.

I take the domain of consciousness, in the relevant phenomenological sense, to include qualia, phenomenal consciousness, experience, what-it's-likeness, awareness, and attention. Whatever phenomena we intend to pick out with these concepts—a fundamental property (Chalmers 1995), an emergent property of brain activity (Crick and Koch 1990), an information-theoretic property (Oizumi et al. 2014), something else entirely, all of the above, or none of the above—it (or they) is (or are) not yet widely ascribed in the unexpected domains. Trewavas (2014), for example, is an exception in claiming that plants may be conscious. It also might turn out that none of the nonhumans should be Literally ascribed psychological capacities because they just don't have them. This would not falsify Literalism, because it is not a form of panpsychism (more precisely, panconsciousness-ism) nor does it entail panpsychism. It does not even claim that all current psychological ascriptions in science to nonhumans are true. They could turn out to be false. The research is new and far from conclusive. But this sort of defeasibility is a feature of any ascription in science, not just those targeted by Literalism.

Moreover, perhaps Nagel (1974) is correct that we do not yet have the concepts or vocabulary that we need to scientifically explain conscious experience or indeed any of "the mentalistic ideas that we apply unproblematically to ourselves and other human beings" (1974: 438 fn. 5). But from the Literalist perspective, we do already have the vocabulary. What we don't know is whether or how scientific findings will alter or eliminate some or all these concepts. We lack widely accepted theories and models that can organize and articulate the pre-theoretic consciousness-related concepts we are using to guide our initial investigations. (The point extends to psychological predicates generally.) What is false is Nagel's implicit assumption that subjective imagination cannot be improved upon by new non-subjective evidence. Untutored subjective imagination may be too limited to grasp the experience of nonhumans—although this may be in part because we don't try very hard. But untutored imagination in general doesn't have a very good track record in terms of understanding the natural world. If we had a model of a variety of consciousness that applied to humans and nonhumans alike (similar to the models presented in Chapter 3), the gap between what I experience and what I can imagine of the experience of a nonhuman may be narrowed to where it is no worse than the gap between what I experience and what I can imagine of your experience. Consciousness concepts could be applied equally unproblematically to nonhumans for good reasons that have nothing to do with whether I, using my subjective imagination alone, find them unproblematic. I may never know exactly what another experiences—whether "the other" is you, your dog, or your dog's fleas. But this residual skepticism need be no worse than external world skepticism—irrefutable from one perspective, unproblematic from another.

Theories of intentionality or mental content are a somewhat different story. To explain, I'll adopt terms from the traditional philosophical framework in which thoughts are analyzed as attitudes towards propositions (sentence-like units composed of conceptual contents).[6] In these

---

[6] I'm not endorsing this analysis, but it is a familiar framework for many philosophers and thus helpful for introductory purposes. In Figdor (2014), I endorse an adverbial view.

terms, Literalism is a theory about attitude ascriptions, not content ascriptions. If fruit flies decide, they make decisions relevant to flies; if bacteria communicate linguistically, what they communicate linguistically will be relevant to bacteria. Naturalistic theories of intentionality or mental content (e.g. Dretske 1988; Millikan 1984, 1989; Fodor 1987) are neutral as to which entities have representations and what contents they acquire. Being able to misrepresent is necessary; being human is not.[7] However, the relevant question here is whether having specifically human content—or, to use the implicitly anthropocentric label, conceptual content—is necessary for having a type of attitude or mental state. I will argue that it is not. Just as many nonhumans really see even if the contents of their visual states are not human, they can have real attitudes even if the objects of their attitudes are not human propositions. In other words, anthropocentrism is no more justified for cognitive states than it is for perceptual ones (assuming these kinds of mental states are importantly distinct).

In general, for a Literalist, many philosophical discussions of "the" mind could simply make explicit the fact that the default subject is the human mind (and often a highly idealized human mind, at that). What the Literalist contests is the presupposition that human cases are the standards for what counts as a real, full-blooded psychological capacity and thus for what counts as a literal use of a psychological predicate. Chapter 2 begins to show why.

---

[7] Roughly, misrepresenting consists in using information individuated in a particular way to refer to or describe circumstances that do not match this content—for example, having a perceptually induced thought that contains the concept COW when what is actually in front of you is a horse. Genuine intentionality at any spatiotemporal scale can require an ability to misrepresent; the empirical problem is figuring out when responses by nonhumans count as mistakes of this sort.

2

Cases

Qualitative Analogy

2.1 General Remarks

In this chapter and Chapter 3, I will discuss a number of different cases in which psychological predicates are being used in unexpected nonhuman domains. These cases are data that any plausible account of the semantics of psychological predicates must explain or else explain away. I believe Literalism is the most plausible account, but I will not begin to make that case in detail until Chapter 4.

The sample cases are divided into those in which the introduction and epistemic justification of psychological predicates in the new domain depend on qualitative analogy and those based on quantitative analogy (in particular, using mathematical models).[1] In this chapter, I discuss psychological predications to plants and bacteria, which involve qualitative analogy. In Chapter 3, I discuss psychological predications to fruit flies and neurons, which involve quantitative analogy. I also sketch some recent theorizing in neuroscience aimed at developing a non-anthropocentric semantics for specific psychological concepts.

The qualitative/quantitative analogy distinction is intended to capture a fundamental difference in the types of scientific evidence we use to establish and justify similarity or resemblance relations, on any analysis of similarity or resemblance. Despite longstanding philosophical qualms about resemblance relations and their resistance to formalization,

---

[1] In related usage, Hendry and Psillos (2006: 145) call the use of mathematical equations in a new domain "analogical transfer", using Bohr's atomic model to illustrate. It is an interesting and open (albeit in this context orthogonal) question whether the same analogical reasoning systems and methods we use for qualitative analogies (e.g. Hesse 1966; Gentner and Markman 1997; Nersessian 2008) are also used in quantitative analogies.

similarity relations are appropriately and adequately specified and widely used in actual scientific contexts (Weisberg 2013: 42–3). The epistemic distinction is critical for understanding the distinct sources of scientific pressure on the semantics of psychological predicates. By "pressure on the semantics" (or "semantic pressure") I mean a strong motivation to reconsider the reference and meaning of an existing term.[2] Such pressures can have sources other than science. As an intuitive example, given recent changes in general social mores in the United States, the legal definition of "parent" is subject to semantic pressure to be extended to include adults who have neither a biological nor an adoptive relationship to a child.[3] A term that is under such pressure is to some extent semantically destabilized, although we can continue to use it even if we don't yet know how its semantics will be resolved or aren't yet fully cognizant of the instability.[4] We need not be aware and in fact are often unaware of the semantic pressures a term is under at a given time, especially if there is a semantic division of labor. In the "parent" example, the openness and instability may not be recognized until a specific legal case is considered in court. The sources of semantic pressure of interest here are scientific practices involving the uses of terms in peer-reviewed contexts to report new discoveries. Literalism and its alternatives are different ways of responding to this pressure.

The problem of scientifically-driven semantic pressure on old predicates is most familiar to philosophers regarding non-psychological cases, such as "gold" and "water". Reference in these cases was settled by giving scientific evidence priority over human perception. In the case of "gold", new scientific evidence and theory led to making the reference and extension of "gold" more precise, as determined by the property of having 79 protons in its atomic nucleus. One might say the rest of us conceded the

---

[2] My main concern is with shifts in reference and extension of psychological predicates, whether one classifies them as general or natural kind terms (which in turn depends on one's theory of natural kinds—an orthogonal issue); they are not singular terms. Further details about reference and meaning (or sense) are discussed in later chapters (especially Chapters 4 and 6); here I am just starting from familiar terrain.

[3] My thanks to an anonymous reviewer for OUP for suggesting this everyday example.

[4] While Literalism is neutral on the issue of semantic holism, clearly instability can spread through a semantic web (e.g. the "parent" instability is directly linked to the pressure on "marriage"). But holism does not entail that every term must be modified if one is, since it does not imply that all terms are destabilized to an equal degree or that all perturbations of the web are equally severe.

issue in the face of this scientific pressure—most of us without incident, others not so much.[5] As will be shown in Chapter 3, mathematical models promise to play the same revisionary role in psychology. In general, they destabilize the reference of predicates denoting qualitatively established similarity classes of behaviors and functions, just as the elements in the periodic table destabilized the reference of count and mass nouns denoting qualitatively established similarity classes of objects and stuffs. Psychological predicates (and behaviors and functions generally) are a new, relatively late, target of this kind of semantic pressure, although much more is at stake when we ask what counts as real cognition rather than real gold or real water.[6]

Both quantitative and qualitative analogies may be used in any single case, for any reasonable way of individuating cases. There may also be historical precedence: the usual case is that qualitative analogies are drawn and quantitative analogies follow (if they do). The puzzles of concept extension via quantitative analogy are no doubt due in part to a lack of historical precedence of qualitative analogy. However, distinguishing them enables me to clarify the different ways they may bring semantic pressure to bear on a predicate. This is important because referential

---

[5] No doubt Captain John Smith was highly discomfited regarding his shipment of pyrite to London: <http://geology.utah.gov/popular/general-geology/rocks-and-minerals/utah-gold/fools-gold/>.

[6] In philosophical debates on the relationship between new reference and old meaning, much attention is paid to the modal status of theoretical identifications, such as the identification of gold with atoms with 79 protons in their nuclei (see Glüer and Pagin 2012 for a helpful review and discussion). For example, does "gold" pick out the same stuff in any possible world (in which that stuff exists) that it picks out in the actual world? This focus takes for granted that we accepted a more precise extension from chemistry that included many items in the old manifest-property stereotype extension but left out others (e.g. fool's gold). That way of responding to the semantic pressure from chemistry was not inevitable—we could have ignored the chemists. My concern is with this prior issue. The Literalist thinks what's happening now with psychological predicates is like what actually happened (not instantaneously!) with "gold" back then. The modal profile of any eventual theoretical identifications is a separate issue. With Mark Wilson (1982, 1985, 2006) I don't think there are strongly context-independent (including temporal context) modal implications of actual uses, even if we had such scientifically established identities in psychology. (Perhaps the logic of the term "identity" generates the problem.) Wilson's position, illustrated with concepts in fundamental physics, may be even more radical than mine, in that on his view (as I understand it) physical concepts have extremely limited factive domains of application ("patches"). Maybe physics is unusually patchy. In any case, his view leaves open how to interpret the semantic relationships of predicates used across patches (the "areas of overlap"), as well as the different ways of individuating patches.

adjustment is more or less easy to resist depending on the kind of evidence used to justify a claim of similarity.

2.2 Background to the Semantic Problem

The traditional semantics of psychological predicates—at least since the emergence of the Manifest Image—is based on qualitative analogy, in particular similarity to and between human beings. The standard inductive inference from the existence of my mind, accessible to me via introspection, to the existence of yours depends on my observation of your behavior; I generalize to the rest of the human species from there. (I discuss this induction further in Chapter 4). I do not adjust the meaning of, say, "pain" when I extend the predicate from myself to you; as Nagel (1974: 435 fn. 5) might put it, I extend it "unproblematically". In principle, of course, this is highly problematic. But as a practical matter, the observed similarity of humans and their behavior is used to justify the presumed stability of meaning across human subjects, whether it does actually justify it or not, and however that meaning and its presumed stability were established to begin with. This dependence on perceived similarity to humans (myself across times or other humans) underlies the received anthropocentric semantics of psychological predicates.

Given this background semantic anthropocentrism, the cases of plant and bacteria cognition discussed in this chapter serve two main purposes. First, they illustrate the general and growing scientifically-based semantic pressure on psychological predicates. The extensions are taking place within the context of what has been described as a Kuhnian (1962) paradigm shift in biology, from a traditional focus on matter to a focus on information as the essential organizing concept for understanding how living systems work (Shapiro 2007: 808). The more specific model-based extensions discussed in Chapter 3 are also made within this broad context. The pressure does not derive from a psychological concept or two used here and there. It stems from a pattern of uses that exhibit the profound theoretical change taking place throughout biology.

Second, they illustrate how a conservative strategy—one that maintains the current anthropocentric semantic standards—can still be used effectively to resist reference revision. Supporters of such strategies see them as ways to interpret the new uses given that predicates just refer to humanlike capacities. They also may adopt, at least at first, traditional defenses of qualitative analogical reasoning in science, such as the fruitfulness of conceptual borrowing for promoting new research (e.g. Trewavas 2003: 355). Of course, borrowed concepts are also useful if they turn out to be literal.

As we will see, such resistance is more difficult in the face of the cases of quantitative analogy discussed in Chapter 3. To anticipate a bit: there need not be any intuitive similarity between human beings and other systems to which the same mathematical model applies, and there need not have been any prior reason to unify humans and other systems on some important dimension. These cases challenge the epistemic priority of the human-centered perspective for establishing relevant similarities and fixing the extensions of psychological categories. As long as we still accept that "in the dimension of describing and explaining the world, science is the measure of all things" (Sellars 1956/1991: §41), it is very difficult to ignore quantitative similarities drawn in contexts of providing scientific explanations and descriptions of natural phenomena, even when those similarities run roughshod over our anthropocentric stereotypes.

Regardless of the reason for revision, however, it might be objected that such a semantic reconsideration is a priori out of the question for commonsense psychological predicates due to how they are defined. This view is articulated within a Wittgensteinian framework by Bennett and Hacker (2003; Bennett et al. 2007), discussed in Chapter 5. However, one might also make this objection independently based on the well-known method of defining theoretical terms articulated and applied to psychology by David Lewis (1970, 1994; see also Churchland 1970; Fodor 1975). On this view, commonsense psychological terms are theoretical terms defined by their roles in a psychological theory, conceived as a set of sentences expressing folk psychological platitudes that everyone knows and everyone knows everyone knows. These natural-language axioms are concatenated into one long conjunction, and each occurrence of a predicate type (e.g. "believes", "is in pain") is replaced by a type of variable to form what is called a Ramsey sentence. In this way each predicate is implicitly defined by and within the theory.

But the objection merely obscures the problem. The Ramsey–Lewis method does not intrinsically limit the predicates to humans. Rather, the platitudes usually cited (when they are) enshrine the qualitative analogies whose epistemic priority is now in question. They are formulated with humans and their actions in mind as the entities to which and the behaviors on the basis of which the predicates are ascribed. The fundamental problem is what determines the right set of platitudes. Even if we agree that the predicates were developed to describe the causes of human behavior (Firn 2004), it does not follow that the causes of human behavior are all that they can describe.

The Ramsey–Lewis method itself is also questionable. It is based on a now-discredited view of scientific theories in which a theory is a set of sentences—lawlike generalizations or axioms—expressed in a science's proprietary vocabulary (Hempel and Oppenheim 1948; Nagel 1961; Craver 2002; Frigg 2006). On this view, theories are language-dependent and the language of a theory is science-dependent. The current leading view, the so-called semantic conception of theories, rejects the idea that vocabularies are proprietary to sciences, just as we reject the idea that mathematics is proprietary to a science. Distinguishing theories from the natural language in which they may be expressed, it holds that theories are models, or assignments of objects to formal structures.[7] The natural language used to describe the theory can be used in different sciences in which the same formal structure is used.

If it is a theory, folk psychology may be a theory in either sense. The point is that there is no good reason to think the Ramsey–Lewis method is the best way to define psychological terms given contemporary views of theories. Instead of treating folk psychology as a set of natural-language platitudes, we can think of it in terms of model-based science, where idealized rational agents represent real-world rational agents (Maibom 2003), and their behavior and capacities are captured in formal structures. From this perspective, the problem raised by quantitative analogy is this: what do we do when a formal model of some aspect of rational agency picked out by a folk psychological predicate turns out to apply to humans and nonhumans alike?

---

[7] The "semantic conception of theories" is so-called because a theory is a model in the logician's sense of a semantics—an assignment of objects to a formal structure (Suppes 1960; van Fraassen 1980). This is not the same as "model" in "model-based science" (e.g. as used by Maibom 2003, cited below), although their relationship is complicated (French and Ladyman 1999; Godfrey-Smith 2006; Halvorson 2012). Nothing here depends on their precise relation. Both the semantic conception of theories and model-based science distinguish formal structures (e.g, logical formulas, mathematical equations) from the interpretation of those structures (e.g. an assignment of objects, a modeler's intended construal), while a language-based ("syntactic") view of theories does not make this distinction. I discuss model-based science further in Chapter 3.

Turning back to qualitative analogy, the Manifest Image's conflation of the psychological (or "person-appropriate") with the human is frequently reinforced by a conceptual chasm drawn between flexible, free humans and inflexible, reflexive everything-else. Intelligence as a concept "is often equated with human intelligence and suppositions of complete freedom of choice (if these exist)" (Trewavas 2003: 17; see also Calvo Garzon 2007: 209). A case of chasm-creation familiar to many philosophers is that of the Sphex (digger) wasp, which has been effectively exploited by Dennett (e.g. 1984) among others.[8] When bringing food back to the nest, the wasp leaves the item at the entrance, goes back inside to check that the nest is safe, and then drags the food in. An early study showed that a wasp would reinspect its nest each time it moved a food item back to the entrance after an experimenter moved it away. It did this over and over again even though it had just reinspected the nest moments before. Behavior that looked intelligent was actually automatic and inflexible.

It turns out that the wasp's behavior is much more complicated and variable (Keijzer 2013). Sphex behavior does not fit the familiar stereotype. In fact, insect behavior does not fit this stereotype. For example, honeybee probiscus extension can be classically conditioned (Bitterman et al. 1983; Boisvert and Sherry 2006) and honeybees appear to have complex navigation strategies that go beyond an egocentric, 'primitive' path integration technique (Wang and Spelke 2002; Giurfa and Menzel 2003; I discuss decision-making by fruit flies in Chapter 3). The Sphex case illustrates, in microcosm, the progression from the Original to the Manifest to the Scientific Image. Within the Manifest Image, we may not observe flexibility in nonhumans because we do not expect to find it.

Both walls of this conceptual chasm are eroding. On the one hand, much human behavior intuitively considered intelligent is not under conscious or deliberative cognitive control; "introspective accounts of

---

[8] As a note of caution, Dennett appears both as foil and ally throughout this book; I discuss his views in detail where appropriate. In many ways we are in complete agreement. From my perspective, Dennett *ought* to be a Literalist and agree that, like the concept of design, psychological properties in nonhuman domains are "as real as it gets" (Dennett 2014: 49). Nevertheless, his most consistent description of his position is in terms of Exsanguinated Properties (Chapter 7), apparently because he is persuaded that the homuncular fallacy is a fallacy. That it is not a fallacy (Chapter 8) does not entail Literalism, but it does eliminate what may be Dennett's primary motivation for this alternative.

the basis of choice should be taken with a grain of salt" (Camerer et al. 2005: 11; Haidt 2001; Kahneman 2011 for a popular audience). Given the roles of affect and automatic processing, its causes are ill understood in terms of the belief–desire–intention framework of folk psychology in philosophy of mind, which in any case is not identical to actual folk psychology (Andrews 2012). On the other hand, assumptions of animal stupidity are being questioned even by researchers who tend to support non-cognitive or "killjoy" explanations of animal behavior (Shettleworth 2010b) rather than cognitive ones (Emery and Clayton 2004; Penn and Povinelli 2013 defend a middle ground). Even language, one of the most cited hallmarks of human cognition, is not immune to this erosion: other species exhibit informative vocal patterns that correspond to phonological regularities in human language (Berent 2013: 324; Harms 2004). New research makes it increasingly implausible to account for species differences by aligning psychological categories with the human category, as the Manifest Image does.

However, the many notable and well-publicized developments in animal research may be considered mere rumbles in an otherwise stable conceptual landscape. In the broad scheme of things, the behavior of a blue jay is familiar and recognizably similar to ours in some ways. These rumbles become tremors when scientists start using psychological terms with respect to beings to which we hardly relate at all, such as plants and unicellular organisms. In presenting some of this research below, I do not intend to provide a comprehensive literature review or weigh evidence for or against the conceptual extensions in each case. I intend to give a sense of the contemporary empirical landscape in which the unexpected uses are occurring. For this reason, I rely on literature reviews by experts who argue for the conceptual extensions and supplement these with a selection of results by distinct researchers (even though the review authors have made significant contributions).

2.3   Qualitative Analogy: Plants

The idea of plant intelligence is not new. Charles Darwin (1880: 573; Trewavas 2007: 231) famously analogized that the tip of the root acts "like the brain of one of the lower animals; the brain being seated within the anterior end of the body, receiving impressions from the sense organs and directing the several movements". Subsequent research has led plant biologists to expand on Darwin's qualitative analogy rather than abandon it.

Researchers have found that plants exhibit many forms of sensitivity and adaptive responsiveness to their environments. These appear analogous to animal behavior that (for many) manifests learning, individuality, choice, and other capacities. As a general example, plants are able to distinguish self from non-self, enabling a plant to avoid competing with itself for resources or to devote more resources to competing with other plants (e.g. Mahall and Callaway 1996). In one study, Falik et al. (2003) created pea plants (*Pisum sativum*) that had two equal roots and two equal shoots ("double plants"). Some of these were split vertically into two equal halves to create physiologically separate "twins". They then planted each double plant (intact or severed) such that each root targeted for measurement had two neighbors, one a self root (intact or severed) and the other an alien root, or else by two alien roots. They found that development was greatest in roots surrounded on both sides by non-self roots, and that most and greatest development was towards roots of different plants regardless of genetic identity. They also found that roots of intact plants grew more towards non-self roots than self roots, whereas roots of severed plants grew equally towards both non-self and (physiologically severed) self roots. They suggest the main mechanism in self/non-self discrimination is physiological coordination, perhaps by means of electrical oscillations that amplify in the presence of neighboring roots, rather than a non-self (allegenetic) recognition mechanism, such as by release of chemicals.

More important for my purposes, a prominent area of recent research has been in investigating the signaling capacities of plants. These include long-distance electrical signaling mechanisms and the roles of molecules that are homologous to neuroreceptors and neurotransmitters in nervous systems.[9] Plant electrophysiologist Jagdis Chandra Bose (1926) found that long-distance, rapid, electrical signaling stimulated leaf movement, and that plants produce continuous, systemic electrical impulses. These findings led Bose to conclude that plants "have an electrochemical pulse, a nervous system, a form of intelligence, and are capable of

---

[9] Homologues are features that are similar due to common ancestry, though they diverge in function (arms and bird wings). Homoplasies are features that are similar in function but have different ancestry (bird and bat wings).

remembering and learning" (Brenner et al. 2006: 414). These and other findings were downplayed for years among plant biologists. One barrier to accepting plant intelligence was the assumption that plants were passive automatons (Trewavas 2003: 2). Rigid, thick cell walls also seemed incompatible with intracellular signaling. Finally, claims about plant spirituality in the popular book *The Secret Life of Plants* (Tompkins and Bird 1973) made it embarrassing to draw plant–animal analogies in serious science (Brenner et al. 2006: 414–15).

Scientific interest in plant intelligence was renewed by Trewavas' (2003) extended review of discoveries of electrical, hydraulic, and chemical signaling capacities that appeared analogous to neural signaling mechanisms. Plants have many of the components found in the animal neuronal system, including voltage-gated channels and plant hormones (in particular, a family of hormones called auxins) that appear to play a cell-to-cell signaling role in plants qualitatively similar to the role of neurochemicals in neuron-to-neuron transmission. Many neurotransmitters, such as glutamate, dopamine, acetylcholine, GABA (γ-aminobutyric acid), and seratonin, are found in plants. A role of these chemicals in signaling in plants has not been established, but the strongest support so far for such a role has been found for glutamate, GABA, and acetylcholine. For example, Lam et al. (1998) found a family of genes encoding for putative ionotropic glutamate receptors in *Arabidopsis*.[10] The gene sequences are similar in structure and have extensive sequence identity to the analogous sequences in animals that code for ionotropic glutamate receptors (GluRs). They also found evidence that the genes code for plant glutamate receptors that regulate light-specific signal

---

[10] *Arabidopsis* is a genus in the mustard family; although they do not specify the species explicitly, *Arabidopsis thaliana* is widely used as a model organism (analogous to mice), including for studying the effects of light on cell elongation (cell wall extension). An ionotropic receptor is a molecule in an ion channel (a gated pathway across a cell membrane) that opens and closes the channel to let ions in and out of the cell. A ligand is a small molecule, such as a neurotransmitter, that binds with a receptor, which then changes shape to allow ion flow. (Reuveny 2013 describes this change in shape for a family of such ion channels in neurons, called GIRK ("G protein-coupled inwardly-rectifying potassium") channels.) In brains, these changes in relative polarization on both sides of neural cell walls, due to ion flows in and out, are the means by which electrical signals are transmitted from neuron to neuron (along with releases of neurotransmitters in synapses between neurons). These signals are (somehow!) the means by which information is processed in the brain.

transduction. In separate experiments, plants treated with a known antagonist to animal GluRs were impaired in inhibiting hypocotyl elongation and in accumulating chlorophyll when exposed to light after being grown in the dark.[11]

These and other findings have recently been consolidated in a new subdiscipline (to skeptics, under a misleading label) of plant neurobiology, whose distinct goal within plant science is "to illuminate the structure of the information network that exists within plants" (Brenner et al. 2006: 417; Baluška et al. 2006; Baluška and Mancuso 2007, 2009). The longstanding analogies between plants and animals are now being conceptualized within the information-processing framework that launched cognitive science. This emerging systems-biology approach to plant physiology is displayed in the interpretation of research results in terms of constructing spatial maps, exhibiting trial-and-error learning, engaging in avoidance behaviors, exhibiting forms of memory, and making decisions regarding when and where to forage for nutrients or what organs to generate or senesce (Trewavas 2003; Brenner et al. 2006: 413). For example, Kelly (1992) tied individual stems with growing tips of the parasitic plant *Cuscuta europaea* (European dodder) to potential hosts that were treated with different levels of nutrient (nitrogen). She found that the parasite stems were more likely to accept potential hosts of high nutrient value and reject hosts of low nutrient value even before taking up food from the host. Kelly describes this as making choices to accept or reject; Trewavas (2004: 15) describes it as active choice influenced by anticipated reward.

Defenders of the new framework argue that such language is useful for reconceptualizing familiar problems, promoting new avenues of research, and conceptualizing accumulating results (Trewavas 2004: 356; Brenner et al. 2006, 2007; Calvo Garzon 2007). The new perspective highlights the importance of investigating plants in their ecological contexts, just as animal researchers recognize the importance of observing animals in theirs. Given this shift of conceptual framework, extending the

---

[11] A receptor antagonist blocks or dampens a response caused by the binding of a molecule (an agonist) to a receptor site. In this case, it appears to dampen light-specific responses to the binding of glutamate to plant GluRs. A hypocotyl is a structure in a seedling between the cotyledon (seed leaf) and radicle (seed root). Inter alia, root growth is impaired if the hypocotyl elongates too much.

Lotka–Volterra model of predator–prey relations to plants suggests itself (Arora and Boer 2006; see Chapter 3 for detailed discussion).

These extended uses remain controversial among plant biologists, although not for a priori semantic reasons. The main concern is that the use of neurobiological language to describe plant physiology "does not add to our understanding" of plants (Alpi et al. 2007; Zink and He 2015). The uses also contradict traditional intuitions about the passivity of plants and their lack of simple criteria of individuation. But we are biased to ascribe mental traits to items that move at dynamical timescales of humans (Carey 1988; Trewavas 2003: 16; Morewedge et al. 2007). Anthropocentrism runs deep. The concept of the intuitive individual as possessor of cognitive states is also under pressure by extended mind and/or dynamical systems theorists in cognitive science (Calvo Garzon and Keijzer 2009; Chemero 2009; Clark 2013). The comparisons Trewavas (2014: 97–103) makes between the behaviors of animals or colonies and plant organs "force us to reconsider our very notion of individuality" (Zink and He 2015: 724).

Nevertheless, the qualitative analogies that are the basis of these extensions do not force us to reconsider the semantics of psychological concepts because they do not provide a non-anthropocentric reason for the claims of similarity. Qualitative analogies between humans and nonhumans have fueled debates in animal cognition for years without raising the underlying issue of whether we are scientifically justified in assuming that the source of the analogy (humans) should also be the standard for what counts as a demonstration of a real cognitive capacity in the targets. A source is not necessarily a standard, and "found first in Xs" does not entail "only Xs really have it". If electrons had orbited around atomic nuclei, they would not have orbited in an inferior way relative to planets. In paradigm cases of qualitative analogy, the resemblance relations are based on observation or visualization, including imagistic reasoning (Nersessian 1992; Gentner and Markman 1997). In psychology, analogy to the human case remains the basis for these extensions. Evolutionary considerations may also suggest homology or homoplasy, but these are also qualitative analogies. Thus, while plant cognition research puts pressure on the semantics of psychological terms, it does not (yet) touch this background anthropocentrism.

Of course, terms initially extended by qualitative analogy can gain quantitative backing later. This is arguably a goal of plant neurophysiology, given Alpi et al.'s (2007: 136) claim that the scientific gain of using the concept of plant neurobiology "will be limited until plant neurobiology is no longer founded on superficial analogies and questionable extrapolations" and instead is grounded in an "intellectually rigorous foundation". Mathematical models of cognitively relevant synaptic mechanisms in brains that also apply to plant signaling mechanisms would fill this bill. For example, Bose and Karmakar (2003) draw parallels between calcium signaling and neural network activity, and propose a model of calcium signaling intended to capture plant learning and memory.

2.4   Qualitative Analogy: Bacteria

Parallel developments are occurring in microbiology regarding the application of cognitive concepts to microorganisms, both prokaryotes and eukaryotes.[12] Amoebas, which are eukaryotes, can anticipate and recall events, and the plasmodium of the slime mold finds minimum-length solutions to mazes (Nakagaki et al. 2000; Saigusa et al. 2008). As Shapiro (2007: 808) puts it: "[O]ur status as the only sentient beings on the planet is dissolving as we learn more about how smart even the smallest living cells can be." For relative brevity, I focus on cognition in bacteria, one of two domains of prokaryotes. *Escherichia coli* is the standard model organism in this area.

The sense of inappropriateness of psychological ascriptions to bacteria is not because intuitively they are passive and not easily individuated, but because intuitively they (along with other unicellular organisms) are too simple. Even within microbiology, until recently "bacteria were generally considered to be little more than 'bags of enzymes', too small to use the complex processes of signal transduction to regulate cellular processes, such as gene expression, not to mention intra- and intercellular communication" (Hellingwerf 2005: 152). For many philosophers, this perspective is illustrated (and perhaps encouraged) by Dretske's (1988) example of bacteria that move towards a gradient. Dretske uses the example to illustrate a form of causal response to the environment that does not merit description in terms of a full-blooded capacity for

---

[12] Eukaryotes and eukaryotic cells are organisms or cells that have internal structure, in particular nuclei and other membrane-bound organelles. Prokaryotes (bacteria and archaea) lack this internal structure.

representation. However, descriptions of bacterial behavior such as "moves towards a gradient" reveal none of the complexity that has prompted explaining it in psychological terms. For starters, bacteria are too short to detect a gradient and cannot move in a straight line for more than a few seconds due to Brownian motion (Ben Jacob et al. 2006: 503–4). They can swim in a direction or tumble randomly. To move towards a gradient, they sample frequently from the environment in which they are swimming; if the food concentration increases, they delay tumbling. The net effect is biased random movement (a biased random walk) towards a higher concentration of food. For once using scare quotes, Ben Jacob et al. (2006: 503) call this frequent sampling of the food level "sniffing". But Auletta (2011: 266–70) argues that *E. coli* behavior is appropriately described in cybernetic terms of information control, and Hazelbauer et al. (2008) suggest it is more accurate to say they track the gradient and respond accordingly.[13] "Moves towards a gradient" skims right by all of this complexity.

The conception of bacteria as bags of chemicals has been abandoned in recent decades to accommodate discoveries of internal functional complexity and coordinated multi-organism (colony) behavior. The development of molecular genetics and genome sequencing techniques applied in microbiology led to the discovery of many regulatory mechanisms in bacteria, including a new class of mechanisms called two-component signaling systems. Two-component signaling systems are a primary means by which bacteria sense and respond to their environments. They are found in some eukaryotes but not in humans or animals, in which a different phosphorylation scheme predominates (Stock et al. 2000: 187–8; Laub 2011: 45).[14]

---

[13] Allen and Bekoff (1999: 146) fall on the conservative side, but the point of their discussion of bacterial behavior is that even if information-processing descriptions are correct, it would not follow that these organisms have conscious experiences.

[14] Signal transduction is "a process by which an external signal is transformed into an entirely different physical or chemical form that affects gene expression or regulatory activity" (Hellingwerf 2005: 152). For example, a photon may be transformed into a phosphorylated protein when a kinase (an enzyme) transfers a phosphoryl group from ATP (adenosine tri-phosphate) to an amino acid in the protein. Phosphorylation (including autophosphorylation) is the addition of a phosphoryl group to a protein or other organic molecule; de-phosphorylation is its removal by a phosphatase (an enzyme). Phosphorylation and de-phosphorylation are primary mechanisms for regulating protein activity. Volkman et al. (2001: 2429) call the heart of signal transduction "the switching of proteins between inactive and active states".

In a two-component system, extracellular stimuli are detected by a sensor protein (a histidine protein kinase or HK, such as CheA) that autophosphorylates and then transfers its phosphoryl group to a response regulating (RR) protein (such as CheY). Two-component signaling pathways can also combine into phosphorelays of multiple phosphoryl transfer events (Stock et al. 2000: 201–3). Hundreds of different kinds of two-component (HK-RR) systems have been identified in prokaryotes, of which each species has a small subset. For example, *E. coli* has thirty HKs and thirty-two RRs (Stock et al. 2000: 188). Bacterial sensing mechanisms also adapt in real-time response to stimuli. In chemotaxis, or movement in response to a chemical-rich environment, this is by means of reversible methylation of proteins at the membrane.[15] A temporal difference in receptor modification via methylation followed by rapid demethylation when a chemical is removed allows for comparison between current and recent past chemical levels (Goy et al. 1977). This comparison process has been characterized by some as form of short-term memory (Taylor 2004; Van Duijn et al. 2006: 160–1) or as information processing in the form of feedback control (Bourret and Stock 2002; Bibikov et al. 2004). Finally, the variety of components, modular construction, and numerous regulatory mechanisms operating on each component in two-component signaling allow for great flexibility and specificity of behavioral response to stimuli. There is also considerable evidence of specificity between component pairings that ensures signal fidelity and hence fidelity of information flow (Stock et al. 2000: 188; Laub 2011: 50–1).

Summarizing these empirical developments, Hellingwerf (2005: 155) suggests that phosphoryl flow in bacteria is analogous to the action potential in brains. These "phosphoneural" networks could give the bacterium "properties associated with intelligent cellular behavior, such as associative memory and learning, and thus with a minimal form of intelligence". The qualitative analogy yields predictions regarding the system's performance that will be testable given more knowledge of the processes, more detailed mathematical modeling, and additional systems-level analysis (Hellingwerf 2005: 156). Finally, viruses (such as the Mu

---

[15] Methylation is the adding of a methyl group to a protein, in this case to the aptly named membrane-bound methyl-accepting chemotaxis proteins (MCPs). Researchers have also found methylation-independent adaptation mechanisms (Bibikov et al. 2004).

bacteriophage) and bacteria transpose their genomes in replication and have versatile and non-random DNA rearrangement mechanisms that help them respond to selective challenges (Toussaint et al. 1994).

These capacities "fit well with a more contemporary view of cells as cognitive entities acting in response to sensory inputs" (Shapiro 2007: 809):

> Comparisons to electronic information systems are useful because they allow us to think concretely and scientifically about complex information processing.... Our digital electronic computing systems are far simpler than the distributed analog processors in living cells...[T]he take-home lesson...is to recognize that bacterial information processing is far more powerful than human technology. (Shapiro 2007: 816)

In other words, the environmental demands on a lectern that one might describe as wanting to stay in the center of the universe (Dennett 1981/ 1997) are not remotely comparable to those facing a single bacterium. A lectern does not behave in any way that would count as an independently motivated explanandum for which a psychological ascription would be an explanans. In sharp contrast, bacteria exhibit controlled behavior that prompts just such ascriptions.

This new conceptual and explanatory framework for microorganisms has also been driven by discoveries at the colony level. Bacteria self-organize into hierarchically structured colonies that adopt cooperative survival strategies (Ben Jacob et al. 2006: 504). For example, they appear capable of cheating behavior, in which genetic mutants that do not contribute to colony welfare have a selective within-group advantage in potential reproductive success. In one study, Velicer et al. (2000) compared *Myxococcus Xanthus* cultures containing deficient clones and their wild-type progenitors. In conditions of starvation, *M. Xanthus* forms fruiting bodies (via intercellular signaling) in which a minority of the individuals in the original population sporulate, i.e. become stress-resistant spores that can germinate when food sources become available again. When a mutant strain defective in sporulation is mixed with a proficient wild-type strain, the defective mutants may sporulate with their same efficiency in pure culture (null hypothesis 1), be completely rescued to achieve wild-type efficiency in mixed culture (null hypothesis 2), or sporulate with greater efficiency in the presence of the wild-type than a neutral wild-type mixed with the original progenitor wild-type (cheating). Using six defective clones in mixed cultures with a single wild-type strain, Velicer et al. found that three produced spores at levels even higher than those predicted by H2. In the most marked case of cheating, a clone that was almost entirely defective in pure culture produced around fifty times more spores than would a neutral wild-type when it was introduced at a level of 1 percent of the mixed culture. In addition, for two cheater clones mixed in nine different frequencies with wild-type progenitors, higher frequencies led to less efficient sporulation, showing evidence that the cheating harmed the mixed groups' performance.

Bacteria may also access environmental information most efficiently when organized as a superorganism or colony (Ben Jacob et al. 2006: 496). As a result, the colony has been reconceived as a single multicellular organism or a social organization (both of which are apt for psychological description).[16] Intracellular self-organization capacities are described as capacities to engage in cooperative behavior, enhance colony identity, make colony-wide decisions, and recognize and identify other colonies, and other group activities. For example, *M. Xanthus* bacteria are considered social microbes; they move together and prey together, surrounding and killing prey by secreting compounds that break down their cell walls (a process known as lysis: e.g. Burnham et al. 1981). Sanchez and Gore (2013) note that models of predator–prey population dynamics, such as the Lotka–Volterra models discussed in Chapter 3, typically assume that neither species evolves—that the timescale of evolution is too long to affect the population dynamics. These timescales can overlap for social microbes, allowing for study of eco-evolutionary feedback loops. Their study with yeast would be a model for other groups with longer temporal scales.

This research presumes that characterizing bacteria in terms of cooperative activity is useful and appropriate (a common metaphor for *M. Xanthus* predatory activity is the wolf pack). Cooperation, in turn, naturally brings in the concept of communication. One common form is called quorum sensing, which regulates behaviors that are effective only in groups (such as sporulation and mating) and that only occur when there is sufficiently high cell density to participate in census-taking

---

[16] Bacteria may behave both collectively, as individuals in groups, and as superorganisms. Huebner (2014) is a recent attempt to provide general principles for when cognition at the group level is plausible. I consider group cognition briefly in Chapter 10, but leave further discussion to future work.

(Bassler 2002). Quorum sensing operates by means of extracellular autoinducers (such as an acylated homoserine lactone or AHL) that diffuse through the cell membrane and accumulate in proportion to cell density until a threshold is reached for stimulating protein binding. There are also anti-quorum sensing strategies ("a form of 'censorship'": Bassler 2002: 424) that can aid such behaviors as helping one species avoid detection by another in mixed populations. Ben Jacob et al. (2004) claim that forms of intracellular chemical signaling should be considered meaningful linguistic communication, with "assignment of contextual meaning to words and sentences (semantics) and conduction of dialogue (pragmatics)" (Ben Jacob et al. 2004: Box 2, 367). They argue that such communication skills enable colonies to develop collective memory, generate common knowledge, learn from experience, recognize other colonies, and organize themselves into multi-colony communities.

2.5 Concluding Remarks

General claims about internal complexity and behavioral flexibility have long been wielded as blunt instruments to justify restricting psychological concepts and intelligence to humans. New discoveries in plant and bacteria behavior and capacities undermine this simple association. It is no longer clear what justifies a sharp conceptual distinction between complex, flexible humans and simple, reflexive non-humans. Cognitive concepts used in connection with plants and unicellular organisms are motivated by discoveries of far greater internal complexity and behavioral flexibility than previously known or anticipated given our "intellectual prejudices" (Shapiro 2007: 215). Because the ascriptions involve beings that are very different from humans and are intuitively far beyond the range of real cognition, they destabilize the semantics of psychological predicates in a way that recent work in animal cognition does not. The claim that Western scrub jays have episodic memory is tame in comparison.

Nevertheless, the justification for these extensions is qualitative analogy, drawn between newly discovered capacities in these domains and human capacities that merit psychological labels. These cases do not seriously threaten the semantic status quo for determining the extensions of psychological predicates, because qualitative analogies have not provided reason to challenge the traditional anthropocentric semantic standards of the predicates. They do not offer a non-anthropocentric perspective on the capacities, as will become more clear in Chapter 3.

In addition, the uses are not clearly interpreted or semantically stable, even among supporters. Despite championing the new conceptual framework for plants, Trewavas (2007) and Brenner et al. (2007) are semantically conciliatory in accepting that the uses may be metaphorical or in some other way not literal. Ben Jacob et al. (2004: 367) at times embrace a metaphorical interpretation of bacterial cognition, whereas Calvo Garzon (2007: 209 fn. iii and 211) argues that neurocomputational features of plants are intended literally. There can even be competing cognitive frameworks. Tauber (2013) notes two different cognitive frameworks for conceptualizing immune functions, a representational model of the self as agent and a presentational dynamical model he traces to J. J. Gibson. Extensions may also be accompanied by unrigorous definitions of cognition in general or other specific capacities. For example, Shapiro (2007: 812) defines cognition as "processes of acquiring and organizing sensory inputs so that they can serve as guides to successful action". This deflationary definition weakens the concept relative to the existing anthropocentric standard to include far more than humans, but without explicitly questioning the legitimacy of the human standard itself. Quantitative analogies raise this question in spades.

3

Cases
Quantitative Analogy

3.1 General Remarks

In Chapter 2, I presented the general problem presented by advances in biological knowledge to our traditional, anthropocentric understanding of psychological predicates and their referents. These cases help to show that the extensions of these terms to new domains in the light of new knowledge are pervasive. They are widely and systematically used and selectively chosen to describe particular empirical discoveries. They also inspire and help frame further research. I also noted, however, that these cases of qualitative analogy do not challenge the anthropocentric standard by which genuine cognition has traditionally been judged. They strain our intuitive ideas of the proper domain of the psychological to a far greater extent than well-known cases of animal cognition. But they do not explicitly raise questions about our intuitive ideas of mindedness itself, and subsequently of our understanding of the reference of psychological predicates. They do not make transparent the problem to which the Anti-Exceptionalist metaphysics within Literalism is an answer. This chapter will make this problem transparent.

Quantitative analogies draw similarity relations based on the use of mathematical models and equations.[1] These cases also contribute to the destabilization of the semantics of psychology, but at this time are less pervasive than the qualitative analogies discussed in Chapter 2. We don't yet have many highly confirmed quantitative models of cognitive capacities,

---

[1] I will be more precise below about mathematical models. To avoid confusion I do not call these "structural" analogies, because the term "structural" has multiple meanings in relevant literatures; for example, see my discussion of theories of metaphor in Chapter 6.

and the model-based extensions of terms are restricted to the few domains in which the models have actually been used. However, the use of quantitative methods throughout biology (not just in population genetics) is increasing as dynamical systems modeling, computational modeling, and network science develop (Bullmore and Sporns 2009, 2012; Baronchelli et al. 2013; Fagan 2016).[2] This includes the uses of models in cognitive neuroscience (e.g. Irvine 2016). We can reasonably expect to see many more cases like the two discussed below as mathematical models of individual human behavior and cognition and human social networks are developed.

That said, the models we do have suffice for my purposes. They show how difficult it will be to preserve the anthropocentrism of our current understanding of the reference of psychological predicates. As we will see in Chapters 5–7, conservative strategies that implicitly deny reference change are still live options, at least in principle. But the metaphysical problem raised by recent scientific discoveries will be out in the open.

The idea that cognition and biology are continuous, which is rooted in Darwinian evolution, has long been defended by some scientists and philosophers (e.g. Maturana and Varela 1980; Beer 2004; Lyon 2006; van Duijn et al. 2006). This claim of psychological continuity across biological taxa gains real bite when the continuity between humans and nonhumans rests on quantitative models of cognitive processing.[3] In what follows, I'll introduce models and modeling with an eye to highlighting the semantic issues that arise when models are extended across intuitively distinct domains. This general consequence of standard modeling practices has not been emphasized or critically examined in the modeling literature in philosophy of science. I will thus set aside standard debates in that literature in order to focus on the impact of model-based scientific practices on the reference of psychological predicates.[4]

---

[2] While my concern is the general trend, the dynamical systems approach in cognitive science (van Gelder 1995; Clark 2008; Samuelson et al. 2015) has already been highlighted independently as an important approach to cognition.

[3] My discussion clearly extends to artificial agents, but I will focus on biological entities through the book. (I consider hypothetical cases involving water and lecterns in Chapter 4.)

[4] Prominent among these debates are how a model can represent a target (e.g., Callender and Cohen 2006; Frigg 2006), the epistemic justification, explanatory role, and nature of simplification, idealization, abstraction, and representation (e.g. Weisberg 2007, 2013; Batterman and Rice 2014; Levy 2015; Levy and Currie 2015), the ontological status of models (Godfrey-Smith 2006; Frigg 2010; Weisberg 2013; Levy 2015), and the relation between mechanistic and model-based explanation (Piccinini and Craver 2011; Chirimuuta 2014).

Following this general discussion, I'll present two cases of model-based extension of psychological terms: decision-making to fruit flies, and expecting and anticipating to neurons. In each case, the model was developed to explain human behavior (or behavior in humans and other animals), and the psychological language is initially introduced in these contexts to describe humans (at least).

In the final section, I'll introduce recent work that aims to modify psychological terms in the light of this and other empirical research.

3.2 Models and Meaning: The Lotka–Volterra Model

Quantitative analogy is not co-extensive with model-based science, but modeling provides clear cases of it and of how it can lead to pressure for semantic revision. Various conceptions of models and modeling has been elaborated recently by a number of philosophers (Hesse 1966; Wimsatt 1987/2007; Giere 1988; Morgan and Morrison 1999; Webb 2001; Humphreys 2002; Godfrey-Smith 2006; Weisberg 2013; Batterman and Rice 2014; Levy 2015; Morrison 2015). A standard (if not universal) view is that models are idealized structures that represent real-world phenomena. They leave out many details of what they represent and may fail to mirror their targets in other ways as well. In some cases the models are physical entities, which may be built in 3D (scale models), pictured in a diagram, or bred to have specific features (animal models). However, models can also be abstract entities that represent real-world phenomena and are described by mathematical equations or other representational means. For example, a simple pendulum is a model—an idealized, non-actual system that represents real-world pendulums. Mathematical equations describe the simple harmonic oscillation (oscillatory behavior) of this imaginary system and the real-world systems it represents.

To go into the further detail necessary for making the semantic problem clear, I will adopt Weisberg's (2013) framework. It is apt for

Weisberg also distinguishes modeling from abstract direct representation (ADR), when mathematical equations describe real-world systems directly, while Irvine (2016) argues that the use of computational templates (e.g. in cognitive neuroscience) minimizes the difference between modeling and ADR. In addition, while my discussion proceeds in a realist vein, those with independent reasons to be committed to anti-realism can reframe it in congenial anti-realist terms.

my purposes because it unbundles the elements of model interpretation and emphasizes the role of modelers' intentions in modeling. On his view (and some others', albeit using different terms), a model has two basic elements: a structure and an interpretation. This applies to concrete models (physical entities, such as an HO scale model train), mathematical models (mathematically described relations between phenomena), and computational models (algorithms describing state transitions).[5] Model structures are specified by model descriptions, which can take the form of words, pictures, equations, diagrams, or computer programs. The interpretation, or construal, fixes what the structure represents and hence what the model is a model of. The construal contains several elements. The assignment sets up relations of denotation to potential target systems, the scope helps identify the aspect of the target on which the theorist focuses, and the two fidelity criteria determine the standards of evaluation of the model (Weisberg 2013: 45). The main feature of a construal that will matter here is the assignment. It is the intended target relationships between the intended target objects. Elaborating on Weisberg's account, the assignment itself has parts—target objects *and* target relations.

The assignment and intended scope together fix which real-world system the model is intended to represent or denote and which aspects of that target matter for the theorist's research. For example, a plastic model airplane has a type of real airplane as its target, but its intended scope is just the real type's basic shape and outer markings, not its internal engineering. The two fidelity criteria (dynamical and representational) set

---

[5] Concrete models are physical objects whose physical features represent those of real-world phenomena (e.g. a plastic model airplane); mathematical models are abstract objects that stand in mathematically described relations between the real-world phenomena they represent (e.g. the predator–prey populations of the Lotka–Volterra model, discussed below); computational models are sets of algorithms that describe the state transitions of computationally defined objects that represent real-world systems (e.g. Schelling's model of non-racist populations that end up racially segregated anyway). These classifications are not canonical; in particular, the relation between computational and mathematical models is not fixed (Humphreys 2002, for example, classifies the Lotka–Volterra model as a computational model). Whether the Temporal Difference model discussed below is mathematical or computational is orthogonal to my argument. In both models I discuss, the equations describe the theorized internal processing of idealized agents, representing real-world agents, with respect to a particular cognitive function inferred from the agents' behavior. The TD model is implemented in an artificial neural network—a set of equations for generating vectors that represent the activity of real neurons in a real brain.

the standards used to evaluate the goodness of fit between the model and the target. The dynamical fidelity criteria fix how much the model's output—the predictions it makes—can vary from real-system output and still count as an adequate representation of this target. The representational fidelity criteria involve how well the model maps onto the target system, depending on which aspects the modeler wants to represent. Thus, the plastic airplane is representationally a good fit to a hobbyist, even though it is not a good fit at all to someone interested in modeling the mechanisms of a real-world airplane. It is not a good fit dynamically to either hobbyist or engineer—one could not reliably predict what would happen to a real airplane based on what one might do with the toy (not unless, for example, the forces that would smash the toy airplane could be reliably measured and scaled up to the forces that would smash the real-world airplane). In the case of toy airplanes, there is no intention to meet any fidelity criteria other than basic shape.[6]

The fidelity criteria block a certain skepticism about similarity. In principle, one can use anything to model anything else. But (to borrow Weisberg's example) while one can use a marble sitting on a table to represent San Francisco Bay, the shape and dynamical aspects of the Bay won't be represented by the marble unless one has stupendously low standards of fidelity. Low standards are in play when we use random objects on a table, like salt and pepper shakers, to represent the players on a basketball team. But low standards are useless in serious scientific research (as opposed to science pedagogy or science communication), where the ability to manipulate, predict, or treat real-world phenomena are paramount concerns.

A key difference between concrete and mathematical models is that the same mathematical structure can be used to represent a much larger range of real-world phenomena *with equal fidelity*. Like a salt shaker, a plastic airplane can represent anything at all given low enough fidelity

---

[6] Note that the noun "model" is sometimes used to pick out just the idealized entity (e.g. the model airplane, the abstract fish population), sometimes—in the case of mathematical models—the equations that describe the idealized entity (e.g. the Lotka–Volterra equations), and sometimes the structure-interpretation package (e.g. the model airplane plus the modeler's intended construal). Also, the mathematical equations that are *model descriptions* to Weisberg and other philosophers are often just called *models* by scientists. As a result, being wholly unambiguous frequently entails being annoyingly pedantic. I will be as clear in the text as seems sufficient to avoid problematic ambiguity.

criteria, but given somewhat higher standards it represents (at best) all real-world airplanes but nothing else. In contrast, mathematical structures are regularly used to model many physically disparate kinds of real-world phenomena—e.g. a pendulum, a spring, a vibrating molecule, the movement of water, etc.—with the same or approximately the same standards of fidelity. In other words, mathematics is as flexible when used in modeling as it is elsewhere in science. This is another reason to note the importance of the interpretation or construal in what constitutes a given model.[7]

A simple but instructive illustration of these features of models is the Lotka–Volterra model. This was initially developed to model the dynamic relationship between the sizes of certain Adriatic fish populations (e.g. sharks and cod):

$$\frac{dV}{dt} = rV - (aV)P, \tag{1}$$

$$\frac{dP}{dt} = b(aV)P - mP. \tag{2}$$

As originally conceived, this model consists of two idealized fish populations and their mathematically described relationship of relative size, plus an assignment specifying two real-world fish populations and a specific real-world relationship between them as the model's targets. More formally, in the *standard construal* of the model, V represents the size of the prey fish population (e.g. cod), P the size of the predator population (e.g. sharks); the parameter r represents the intrinsic growth rate of the prey population, m the intrinsic mortality rate of the predators, a the capture rate of prey, and b the birth rate of predators. Equation (1) equates the change in prey fish population size over time to the difference between its intrinsic growth rate and the rate at which prey fish are captured by predators. Equation (2) equates the change of predator fish population size over time to the difference between the

---

[7] On Batterman and Rice's (2014) minimal-models account of the explanatory power of mathematical models, the equations define *universality classes* across domains and scales in which many details of systems whose behavior satisfies the equations are deliberately left out. Explanatory power lies in showing why these details are irrelevant.

birth rate of predators and the predator mortality rate.[8] The equations are coupled by the second term in the first equation and the first term in the second equation: the conversion of dead prey into baby predators (in Eq. (2)) is a function of the capture of prey by predators (in Eq. (1)). This model captures the empirical fact (observed by Volterra's son-in-law) that less commercial fishing in the Adriatic during World War I affected the fish populations differently. For a time, the predator population increased relative to the prey population. The model captures the related fluctuations—out-of-phase oscillation—of the sizes of the two populations.

This simple model is sufficient to make the semantic issue of interest clear. The terms used to interpret the variables—"Let **V** stand for . . .", etc.—introduce semantic, specifically referential, elements into the model. For example, the labeling of **P** and **V** in terms of "predator population" and "prey population" express the modeler's intention to pick out a known real-world relationship between real sharks and real cod. Some cod are captured, killed, and eaten by sharks, some of which survive to reproduce as a result. The equations describe this dynamic relationship between the sizes of the populations, given these other relationships between the members of the populations. It is in virtue of these referential terms that equations that describe an idealized relationship between idealized fish populations represent a real-world relationship between real-world fish populations. More precisely, the mathematical model is a model of fish because (i) a modeler intends it to be about fish, (ii) she expresses her intention by saying "Let X stand for fish", (iii) what it is to be a fish is a settled matter, and (iv) "fish" refers to fish. The same goes, *mutatis mutandis*, for any referential term used to interpret the symbols in the mathematical structure. Note that the reference of "fish" in (iv) relies on the settled nature of fishhood in (iii); in this unproblematic case, semantics and metaphysics quietly slip in together.[9]

---

[8] When values are assigned to the equations' parameters, the result is an instantiated model (Orzack and Sober 1993) or an instantiated model description (Weisberg 2013: 37). **P** and **V** are dependent variables of this assignment of values to the parameters: roughly speaking, the relative population sizes expand and contract in a way that depends on how fast or slow one sets the growth, death, capture, and birth rates.

[9] Another way to express (i) and (ii) is to say her use of "fish" expresses her concept FISH, which is part of the content of her intention (a mental representation). This difference won't matter. I'll discuss concepts further in Chapter 6.

Now suppose we have a well-regarded mathematical model, like the Lotka–Volterra model. The semantic problem arises (when it does) when we extend that model to new target objects that have no obvious resemblance to the first, or (what amounts to the same thing) when the real-world relationships specified by the terms used in the original assignment now appear to include a heterogeneous set of objects from a pre-theoretic, often qualitatively-based, perspective.[10] The assignment of a new type of target object puts pressure on a relational predicate's semantics if the new assignment's objects are not within the intuitive extension of the predicate used to pick out the intended target relation.

Some pressures are so intuitively inconsequential that reference revision is easily, even reflexively, deflected. No one cares if V stands for a population of prey fish other than cod. Others are not so easy to ignore. As noted above, the Lotka–Volterra model is usually characterized non-ichthyocentrically as a model of predator–prey relations, even though P and V originally represented fish populations. That was Volterra's original intended assignment of objects. The model was later used to represent relations between population sizes of foxes and rabbits, and wolves and moose. Its customary description as a model of predator–prey dynamics permits (or even enabled) this change of assignment from fish to land mammals to slip in under the radar. We construe the model as being about the relative sizes of *animal populations*, but still related by *predation*.[11] The change in assignment creates no ripples because foxes and rabbits, or wolves and moose, already stand in the extension of the relation denoted by "predation". The notable differences between these predator–prey pairs and sharks and cod are implicitly judged irrelevant (or unproblematic).

This presumed stability of predicate reference across new assignments of objects is strongly undermined when P represents the labor share of national income and V the employment or jobs rate (Goodwin 1967), or when P represents capillary tip growth and V levels of chemoattractant

---

[10] Obviously, if any change in any element in a model entails a new model, then this will be a different but overlapping model. In scientific practice, models are not individuated this finely—the Lotka–Volterra model is an example. See further discussion in the text.

[11] The corresponding growth relation is that of converting prey into baby predators—a relation with no name comprised of nutrition, metabolism, sexual relations, incubation, and birth. For simplicity, I focus on predation in the text.

in wounds (Pettet et al. 2000).[12] When the structure is reinterpreted with an assignment of objects that doesn't fall within the extension of the assignment's relations from a pre-modeling perspective, the model creates a semantic IOU for the relational predicates.

What happens when structural similarity cuts across the traditional extensions of predicates used in an interpretation? How much can a predicate's extension be stretched without snapping into homonymy? In this case, the equations might be reconstrued as denoting competition or symbiosis, but these relation assignments lack the theoretically useful specificity of the original target relation. Alternatively, one might ignore the similarity that has been revealed and declare the relations in each model to be of predation, income distribution, and angiogenesis (new blood vessel growth), respectively. But the utility of a structure across different domains is a positive feature in science that displays unifying explanatory power (as, e.g., Batterman and Rice 2014 emphasize).

Note that these new assignments reveal *at least* a formal similarity between certain kinds of interactions occurring in intuitively (qualitatively) distinct ecosystems. Structures can provide common ground that blocks linguistic incommensurability (Godfrey-Smith 2006: 739). They can help us see old relational concepts in a new light. So the fact that an initial similarity is formal does not imply that it is trivial or that there is nothing more to be discovered. To the contrary, it suggests further investigation. Not every formal similarity will be intuitively interesting, but some will—and what counts as interesting shifts when a formal similarity is found.

Weisberg (2013: 77–8) briefly discusses Goodwin's use of the Lotka–Volterra structure in a different model—different, on his view, because "the entirety of the assignment changes". While he notes that Goodwin "was prepared to take the analogy of predation quite seriously", he also considers Goodwin's use "a bit of an unusual, extreme case, in which a mathematical structure is completely reinterpreted in a new

---

[12] A chemoattractant is a chemical agent that induces an organism or cell to move towards it. This movement is called chemotaxis. Regarding referential stability, I consider natural language predicate reference *always* disposed to change, but usually sufficiently stable (making sufficiently strong manifestations of this disposition unexpected and at times disturbing). See also Chapter 2 fn. 12.

domain". Actually, it is neither extreme, nor unusual, nor unexpected for mathematical models—flexibility is, after all, a major benefit of mathematical models over concrete models. The judgment that the fox–rabbit object assignment is not an "entire" change of assignment, while the wages–jobs assignment is, presumably lies behind Weisberg's claim that the latter are related in a way *analogous* to predation. But declaring that the relation in Goodwin's unexpected use is *analogous* to predation settles the semantic problem by fiat; it may be a version of the Metaphor view (see Chapter 6.) Why think these are analogous relations rather than two cases of the real thing?

From a pre-modeling standpoint, Weisberg is presumably correct that the change in construal from sharks and cod to foxes and rabbits is *not* an "entire" reinterpretation. In both cases, the shifting of energy resources goes via easily recognizable activities of killing, sex, and birth. These activities, and thus the extensions of the predicates referring to them, do not include the activities and events behind shifts in financial resources from job creators (or jobs) to job occupiers (or wages), or shifts in chemical resources from a wound space to budding blood vessels. From this standpoint, one has provided an entire reinterpretation if the predicates in the original construal do not already include the new objects in their extensions. But this way of drawing a distinction between entire and partial reinterpretations takes for granted the correctness of the pre-modeling standpoint towards the predicate extensions.[13] It denies the possibility of updating the extensions in the light of new scientific evidence. This is problematic. When we find a formal similarity, we have an important scientific reason to reconsider the pre-modeling perspective on the predicates.

In this way, standard modeling practices in science can provide strong reasons to question intuitively based categories of properties and relations. From a semantic perspective, when should we generalize across new assignments to include unexpected objects and their relations in the extensions of old predicates, or exclude ones that were included previously? From a metaphysical perspective, what properties and relations do the old predicates denote anyway? From an epistemic perspective, what justifies continuing to slice and dice nature according to our pre-theoretic

---

[13] It may be a reflection of what Weisberg (2013: 68–9) calls the folk ontology of the model. It is certainly a reflection of the ontology that the modeler brings to the table.

lights if we have strong evidence of unity across pre-theoretically distinct domains? These questions transpose into the key of modeling the same issues that arose when the empirical and theoretical developments of atomism and chemistry put semantic pressure on the old reference of "gold". And we know how that one turned out.

On his part, Goodwin explicitly felt more was going on in his use of the Lotka–Volterra equations in economics:

> To some extent the similarity is purely formal, but not entirely so. It has long seemed to me that Volterra's problem of the symbiosis of two populations— partly complementary, partly hostile—is helpful in the understanding... of capitalism.   (1967; quoted in Vadasz 2007: 12)

In other words, behind Goodwin's construal was an insight that the kinds of relationships captured in the equations are not limited to animals and life and death. One may even agree with Weisberg (as I interpret him) that there is a *family* of Lotka–Volterra-based models of predation that are reasonably considered distinct from the economics family at a usefully precise level of description, and from the angiogenesis family if there is one. The problem can then just be reformulated in terms of what determines the boundary of a family and when a model belongs to that family.[14]

For example, what about the use of the Lotka–Volterra equations for plant populations (Arora and Boer 2006)? Patterns of interactions between plants and even plant tissues have previously been characterized in terms of competition and coordination (Trewavas 2004: 353). Is this

---

[14] In general, individuating models is tricky. For example, Weisberg (2013: 37) notes that more or less precise values assigned to the parameters of a model description specify more or fewer models; if the parameters are assigned completely precise values, the instantiated model description specifies a single model. On the other hand, a model is a combination of a structure and a scientist's interpretation or construal of that structure (2013: 24, 34, 39 passim). It follows that a precisely instantiated model description specifies a single model only if just one construal is possible for that precisely instantiated description. That is implausibly fine-grained for actual modeling practices. Precisely valued instantiated Lotka–Volterra equations can be construed as being about sharks and cod, or foxes and rabbits, or predators and prey, or wages and jobs, or capillary tips and chemoattractant (and so on). The same parameter value assignments will yield the same values of P and V, but this does not tell us how the construals are related and so how many models we have. In short, once one has more than one construal of the same structure, the question of how the construals are related arises and there is no one way to answer it. This is not a criticism of Weisberg (who directs badly needed attention to these issues) but a cautionary note about model individuation.

model within the predation family? Is the idea of predator plants too reminiscent of *Little Shop of Horrors* to be taken seriously, or does our familiarity with Venus fly traps justify the idea that this is genuine predation and not a counterfeit? The model itself provides an important *independent* reason to think plant populations stand in relations of real predation, not mere analogues of predation: their relative size fluctuations satisfy these equations. Their failure to move around like animals or otherwise satisfy pre-modeling stereotypes of "predator" needn't prevent them from engaging in the same, not analogous, behavior. The metaphysical difference between a real relation and a merely analogous relation is not determined by whether one extension of a predicate feels natural and the other doesn't. A feeling of naturalness may explain our initial response to unexpected new cases, but it begs the question to consider that feeling a sure sign of a predicate's proper extension.

In this way scientific modeling practices yield the same potential for reference change for relations and properties that we faced when matter was subject to reclassification in the light of developments in physics and chemistry. Now we are reclassifying relations rather than material stuff, and we are doing so using mathematical structures, not numbers of subatomic particles. As we will see below, the cases of quantitative analogy involving cognitive models applied to nonhumans are very much like the plant example.[15] A model of decision-making developed for humans is applied to fruit flies, and a model of classical conditioning (or reinforcement learning) used across a wide range of human and nonhuman animals is applied to neurons (as well as microorganisms: Mitchell et al. 2009). Unlike qualitative analogy, quantitative analogy provides us with a strong new epistemic reason for revising our traditional understanding of the real extensions of psychological predicates. Because of the flexibility of models and standard modeling practices, quantitative analogy involving cognitive

---

[15] Quantitative analogies can rest on prior qualitative analogies. Garson (2002) describes Edgar Adrian's 1928 discovery in neural electrophysiology of structure in nerve impulses, which "provided the material for analogies between transmission of messages in artificial systems of communication and the transmission of impulses in the nervous system, thereby providing a rationale for the extension of the concept of information from artificial systems of communication to include some biological systems". Alpi et al.'s (2007) call for providing plant neurobiology with an "intellectually rigorous foundation" that would replace "superficial analogies and questionable extrapolations" is plausibly interpreted as a call for quantitative models.

models directly challenges our traditional anthropocentric semantics for psychological terms.

Of course, when the terms under pressure are psychological, the problem is up close and personal. It is one thing to comprehend a similarity between sharks, labor share of national income, and capillary tips using a model usually thought of as being in part about predators. It is another to comprehend a similarity between humans and nonhumans because of shared structure in patterns of behavior. In the case of humans, we infer without fanfare from these patterns of behavior to psychological capacities. In later chapters, we will encounter strong resistance to the claims that these common structures justify making the same inferences to psychological capacities in nonhumans, and that, in consequence, we should interpret the psychological predicates as being used literally, with the same reference, across the relevant nonhuman and human domains. Yet part of what makes Literalism difficult to resist is the importance of modeling in contemporary science coupled with the post-Galilean epistemic priority of science in explaining natural phenomena.

3.3 Quantitative Analogy: Fruit Flies and the DD Model

In 1978, cognitive psychologist Roger Ratcliff introduced a theory of decision-making that came to be known as the drift-diffusion model (DDM) or diffusion decision model.[16] In this model of the cognitive processes involved in quick (1 or 2 second) decisions, the subject accumulates evidence until a response criterion is reached and a decision is made. Considered "one of the cornerstones of modern psychology" (Milosavljevic et al. 2010), it is used for humans (young and old, healthy and impaired) and nonhumans alike in a wide variety of experimental contexts (Ratcliff and McKoon 2008: 20–2; Milosavljevic et al. 2010; Ratcliff et al. 2016). Sequential sampling (that is, evidence accumulation) decision-making models, of which the DDM is an exemplar, account well for the basic speed–accuracy tradeoff in single-stage decisions

---

[16] The name comes from the fact that the additive accumulation of evidence (drift) exhibits variations in the rate of accumulation that can be represented as a diffusion process (i.e. Brownian motion). See Palmer et al. (2005).

between two alternatives (Ratcliff and Smith 2004; Ratcliff et al. 2016).[17] Such models posit a common process of evidence accumulation that links decision speed and decision accuracy. In sequential sampling, a stimulus is repeatedly sampled, generating repeated and varying representations of it. These are added to accumulate evidence that reaches one of two decision boundaries. Is it a house or a car? Is it the same figure again or a different one? Is it a word or a nonword? Is the direction of coherent motion to the left or the right? Perceptual, lexical, category, and other decisions are among the sorts of decisions to which the model has been applied. Other types, including value-based decisions, are being explored.

Ratcliff (1978) modeled this type of decision-making process as a continuous random walk—a semi-random process—in which evidence for a decision is accumulated (summed) over time until a threshold (or boundary) for a decision is reached and a response indicating that decision is made. The model generates predicted levels and distributions of speed and accuracy outcomes for different levels of difficulty (among other manipulations). If the actual data from experimental subjects follows the same curves within standard tolerances, the model fits or accommodates the data.[18] For example, human subjects presented with visual stimuli of varying degrees of degradation or noise are instructed to decide as quickly as possible or as accurately as possible (or both) whether a test stimulus matches or does not match a sample stimulus or a representation of it in memory. The drift rate is the rate of accumulation of evidence. Manipulating the quality of the stimulus modulates the drift rate. For example, conditions with less noise have higher drift rates. Different instructions modulate the criteria for how much information is required for a decision, or where the decision boundary is set.

---

[17] Ratcliff et al. (2016: Fig. 1) shows relationships between sequential sampling models of decision-making that vary in ways that do not change the construal of the model (or the members of the model family)—for example, whether one assumes one or two structures for accumulating evidence, or whether the amount of evidence required to reach the decision threshold gets smaller as the time to make the decision increases or else stays constant. These are variations on the same mathematical theme.

[18] This description is intended to make basic modeling practices accessible to those who are unfamiliar with them. It ignores a plethora of theoretical, conceptual, and practical issues involved in curve-fitting. These issues in philosophy of science (specifically philosophy of statistics) are by no means specific to this discussion and can be raised for any modeling practice.

In general, increasing difficulty increases the mean response times and decreases accuracy.[19]

In more formal terms, the standard model divides the reaction or response time (RT) into three components: u (encoding time) + d (decision time) + w (response output time). Manipulating the drift rate and decision boundaries yields different values for these variables. The model distinguishes non-decision processing involved in making information of differing quality available for a decision from the decision process itself. The non-decision component of the RT is $u + w$ (for visualization, see Ratcliff and McKoon 2008: 876, Fig. 1, bottom panel). u in turn represents components of processing, such as stimulus encoding and representation, that also can be modeled, with the results being used to provide values for the DDM.[20]

DasGupta et al. (2014) used a simple drift-diffusion model (from Palmer et al. 2005) to model the responses of FoxP mutant and wild-type fruit flies (*Drosophila melanogaster*) in a two-choice odor-discrimination decision-making task. The design of this study drew on a number of independent scientific milestones in addition to the drift-diffusion model: classical conditioning, Shadlen and Newsome's (1996) experimental design for decision-making using macaque monkeys, research on the FoxP genetic mutation in humans and other species, and genetic engineering to create FoxP mutant fruit flies. The FoxP gene subfamily has been linked to language and speech disorders in humans, although its functional role has recently been broadened to include learning skilled motor coordination (Mendoza et al. 2014). DasGupta et al. found that $FoxP^{5-SZ-3955}$ mutants took more time to make decisions of at least the same level of accuracy as wild-type flies, indicating a role of the $FoxP^{5-SZ-3955}$ gene in integrating evidence. This result explains the study's significance and its appearance in a major scientific journal. The study did not aim to show that fruit flies make decisions. The goal was

---

[19] The means and distributions of the outcomes also change in regular ways (Ratcliff et al. 2016); while important for showing the ability of the model to account for the data, this detail is not relevant here. I should also note that the evidence accumulation aspect of the model may well be revised to incorporate recent predictive coding theories of neural function (e.g. Hesselmann et al. 2010).

[20] Ratcliff (1978: 60) suggested a metaphor of a tuning fork's resonance to suggest the nature of the evidence comparison process: an item in memory and a probe item are like tuning forks that come to vibrate at the same frequency. He explicitly omits the metaphor from the model itself.

to see if the genetic mutation made a difference to their decision-making, as measured by differences in speed of response and accuracy by mutant and wild-type flies at the same levels of difficulty.

They first conditioned the flies to avoid a certain odor concentration of a chemical (4-methylcyclohexanol). In experimental trials, odor concentrations were manipulated to make them easier or harder to detect. The decision involved the flies' determining whether the odor concentration in experimental trials reached the level they had been trained to avoid. The closer the experimental concentration to the concentration they were trained to avoid, the more difficult the decision. Reaction time (RT) was quantified as the time between entry into and exit from a specific decision zone within the training and test apparatus, a clear plastic chamber with odor inlet ports and outflow vents.

The DDM they used to fit their data was described by these equations:

$$T(x) = \frac{A}{kx} \tanh(Akx) + t_{residual} \tag{1}$$

$$F(x) = F_{max} \frac{1}{1 + e^{-2Akx}} \tag{2}$$

(DasGupta et al. 2014: Supplementary Materials p. 3; see also Palmer et al. 2005)

$T(x)$ is the reaction time (the "chronometric function", i.e. reaction time as a function of stimulus strength) and $F(x)$ the fraction of correct choices (the "psychometric function", i.e. accuracy as a function of stimulus strength). The equations relate these two functions: both are functions of x, the odor concentration ratios (a.k.a. stimulus strength) manipulated to provide lower- or higher-quality evidence. Manipulating x in turn raises and lowers the drift rate k, the rate of evidence accumulation. A is the decision boundary (i.e. match or non-match, hence avoid or not). F_{max} is the empirically determined fraction of correct choices in the easiest odor discrimination task. Note that the RT (Eq. (1)) is divided into two random variables, decision time (the first term) and residual time (labeled $t_{residual}$), following Palmer et al. (2005); the latter includes non-decision sensory and motor processes, as well as the time to move into the decision zone in the test apparatus (held constant at about 1.5 seconds). Accuracy predictions were made by fitting actual RT data into Eq. (1) and using the resulting values for A and k (at odor concentration level x) in Eq. (2).

The drift-diffusion model captured the relation demonstrated by the flies between difficulty of decision (from easy- to hard-to-detect differences in odor concentration) and performance (the fraction of correct choices, the speed of the decision), just as it does with humans in analogous (but non-odor-discriminating) decision-making tasks. Both types of flies showed similar difficulty-dependent increases in reaction time and error rate, but the FoxP mutants had significantly longer reaction times than wild-type flies in conditions with hard-to-detect differences in odor concentration (see DasGupta et al. 2014: Fig. S3 B Supplementary Materials).

DasGupta et al. consider a number of different explanations of this difference in performance: "the abilities to learn from shock reinforcement, walk to and from the odor interface, detect olfactory cues, and decide". They rule out the first three due to other experimental results. For example, in tests of sensory processing (detecting and processing olfactory cues), mutant and wild-type flies displayed the same odor sensitivity in tests. They then remark:

Where mutant and wild type flies clearly differed was in the dependence of reaction time on stimulus contrast: In mutants, narrowing the odor concentration difference caused disproportionate increases in reaction time.... A drift-diffusion model identified two changes that can account for this phenotype: a 38% drop in drift rate and a—perhaps compensatory—increase in the height of the response bound. The reduction in drift rate suggests that *FoxP* mutants are impaired in the accumulation and/or retention of sensory information in the buildup to a choice.   (DasGupta et al. 2014: 902–3)

Given its fit to both human and fruit fly data, the model helps justify the ascription of decision-related component cognitive processes posited by the model (e.g. evidence accumulation) to the intended target populations of decision-makers. This change of object assignment in a decision-making model appears parallel to the change of object assignment from fish to land mammals in the Lotka–Volterra model. We have no more reason to abandon the standard construal of the DDM as a model of two-choice decision-making than we do to abandon the standard construal of the Lotka–Volterra model as a model of predator–prey relations. Humans and fruit flies appear relevantly similar in this respect. So when DasGupta et al. write that "FoxP mutants take longer than wild-type flies to form decisions of similar or reduced accuracy in low contrast conditions", it is reasonable to interpret them as expressing the following: FoxP mutants take longer than wild-type flies to form decisions of similar or reduced accuracy in low contrast conditions.

3.4 Quantitative Analogy: Neurons and the TD Model

Decisions modeled by the DDM are usually (but not always) single-choice, yes or no, match or non-match, choices, rather than more complicated processes involving preferences and assigning values to different options. In economic models of the latter type of decision, rational decision-making involves constrained utility maximization. Decisions are made by assessing the relative value of various available actions, and actions are assigned values based on the estimated gain from doing that action and the probability that the gain will be realized. To act rationally is to decide to do and then do the highest value option.

Platt and Glimcher (1999) suggested that the same classical decision-theoretic framework used to study human choice should replace reflex models of sensory-motor processing. They found that changes in expected gain from a given action correlated with changes in inter-parietal neural activity of macaques when the monkeys chose between actions (as measured by shifts in eye-gaze) to which they could assign different outcome values. Both macaques and neurons "behaved as if they had knowledge of the gains associated with different actions" (Platt and Glimcher 1999: 237). The resulting research area, called neuroeconomics, is controversial, not least due to the extension of psychological terms used in the standard construals of these decision-making models to neurons. (One might argue that the neurons encode the monkey's knowledge, but do not have knowledge.)

However, like the Lotka–Volterra equations, economic decision-making models are behavioral and articulated at the population-level. The former do not tell us how an individual shark stalks its prey or the rate at which it converts prey into baby sharks. Similarly, in classical economic theory, how individuals acquire and modify their utilities—the cognitive processing that results in value assignments or a decision—is not modeled. The decision variables need not even be literally true of individuals for the model to yield reliable behavioral predictions

(Camerer et al. 2005: 9–10; Camerer 2008). To borrow the behaviorist metaphor, the black box is left closed. Even if variation in inter-parietal neural activity correlates with macaque behavior expressing different value assignments, how the macaque assigns value remains inside the black box (maybe forever).

In contrast to this perspective in neuroeconomics, the following cognitive model, also used for neurons, captures an important type of individual learning process.[21]

In the 1980s and subsequent work, Richard Sutton and Andrew Barto developed the temporal difference (TD) model of reinforcement learning or classical conditioning (Sutton and Barto 1981, 1998; Sutton 1988; Barto 1995). They also indicated the intended scope of their model: "[W]e desire to demonstrate that processing of the proposed complexity is clearly possible at the cellular or simple network level" (Sutton and Barto 1981: 136).

The TD model is an extension of the Rescorla–Wagner learning model, a fundamental learning theory developed to account for many of the main features of classical conditioning in animals (Ludvig et al. 2012; see also Seymour et al. 2004 for its application to humans). The TD model accounts for many empirically known features of classical conditioning. It was inspired by behavioral data from animal learning to make predictions, and has been used to model honeybee foraging as well as more familiar cases of classical conditioning. Below we will see its application to neurons (Schultz et al. 1997; Suri and Schultz 1999: 872). Note that classical conditioning is a cognitive process despite its initial association with behaviorism (Rescorla and Wagner 1972; Rescorla 1988). I will focus here on the quantitative similarity established between neurons and other adaptive systems by means of the TD model.

The TD model posits "neuronlike adaptive elements that can behave as single unit analogs of associative conditioning" (Sutton and Barto 1981: 135).[22] One of their explicit goals was to show "the cellular

---

[21] Camerer (2008: 416–17) cites the TD model as a model of the stable preference or choice patterns that are taken as the starting point for classical economic theory. Many economists remain agnostic about the psychological (and neural) reality of utilities, but this controversy is not important for my discussion.

[22] These are also described as "adaptive elements out of which adaptive systems can presumably be constructed" (Sutton and Barto 1981: 144 fn. 3). This perspective undermines the status of the homuncular fallacy as a fallacy (see Chapter 8).

plausibility" of a learning rule, in contrast with a standard view of neurons as input–output switches in early connectionist modeling. This standard view is still the norm in many presentations of connectionism. However, as Sutton and Barto (1981: 136) remark, "Although one of the most important aspects of model building is simplification, the lack of significant progress in adaptive network theory, together with the high complexity of cellular and synaptic machinery, suggests that these idealizations leave out some mechanisms that are essential for producing sophisticated adaptive behavior." In modeling terms, adaptive elements are idealized learners of how to predict future reward that represent real-world learners, just as idealized harmonic oscillators represent real-world oscillators and their behavior. The TD equations describe an adaptive element's process of learning to predict future reward, just as the equations of harmonic motion describe oscillatory behavior. It so happens that neurons are in the class of adaptive elements, although it took a few decades to obtain that evidence.

An adaptive element learns when its expectations are not met. In classical conditioning terms, the adaptive element uses the conditioned stimulus (CS) to predict or anticipate the upcoming unconditioned stimulus (UCS). Consider the classical Pavlovian case of a dog that hears a bell (the CS) and then is presented with food (the UCS), to which it salivates (the unconditioned response, or UCR). After a number of CS–UCS pairings, the ringing bell (CS) triggers salivating (the conditioned response or CR, which is very similar to the UCR). But not just any CS–UCS pairing results in conditioning. If the CS and UCS are presented simultaneously, no or very poor learning will occur. The bell will not trigger salivating. Also, if one conditioned stimulus (CS1) is paired simultaneously with another conditioned stimulus (CS2, say, a flashing light) after the CS1 has been effectively used in conditioning the CR, the CS2 will have no effect. Behaviorism cannot explain these features of conditioning, since in all cases there is pairing of the CS and UCS (or CS1 and CS2). The missing feature is that the CS–UCS (or CS2–CS1) pairing must also be appropriately temporally related for learning to occur. That is why classical conditioning is now described as the process of learning to predict the future (Ludvig et al. 2012) or of learning a predictive relationship (Sutton 1988). It involves keeping track of temporal relationships and managing expectations of future events. The CS is the earliest and most reliable non-redundant predictor of the UCS. The dog's salivating at the bell is anticipatory behavior. It indicates its anticipating or expecting food at a predicted future time.

The TD model models learning as a prediction-generating and revising process. It breaks the learning of the temporally mediated CS–UCS relationship into a series of steps that each involve gently nudging upwards the associative strength between the CS and UCS. A mechanism linking the learned prediction and consequent anticipatory behavior— e.g. salivating—is not part of the model itself, just as motor processing leading to action after a decision is not part of the drift-diffusion model of decision-making. Salivating is an external indicator of learned anticipation. The equations spell out how the learning process occurs.

Learning is driven by prediction errors. At a given time step t-1, the system predicts both its reinforcement at t and its prediction at t of subsequent reinforcement (which is the discounted sum of all future-time-step reinforcement, up to the time of expected reward onset). Then, at t, it compares the predicted prediction value with the prediction of subsequent reinforcement it actually makes at t, yielding the temporal difference (if there is one). The difference between this term and the actual reinforcement at t is the prediction error. The association weights are incrementally updated by this error term (if there is one). For example, a dog learns to associate a bell ring at t with food reward at t+1 by incrementally updating its estimate of what it should expect at each intervening time step in the temporal interval between t and t+1. Incremental updating continues until it has honed in on the correct temporal relationship between bell ring and reward presentation. After learning, it should predict, and expect, no reward at moment t+.5. Learning occurs whenever there is a change in prediction, not only when the UCS is received or omitted (Ludvig et al. 2012). In connectionist terms, the dog's updated estimates are represented in incremental adjustments to the weights of the connections between neurons.[23]

---

[23] At the neural level, the neuron's updated estimates are inferred from its firing pattern; the mechanisms in the neuron are not known. An artificial neural network (connectionist network) is an idealized representation of a real neural network and its state transitions (the passing of activation between neurons). Associations in neural networks are made by increasing the weights between nodes, so that when one is activated the other is more likely to fire as well. The equations presented below describe state transitions of this idealized representation of real neural populations, such that when the network has learned, its weights have been adjusted so as to transform input into the correct output.

Barto (1995: 13) likens the learning process to "the blind being led by the slightly less blind".

The basic TD model description consists of linked equations for generating the reward (or US) prediction and the prediction error as well as for generating the weights assigned to stimuli (what should be paid attention to) and the eligibility for modification of an association (its plasticity). These equations govern the gradual increase in association strength between a stimulus presented at a given time and a reward that is expected to occur at a future time. A simplified version (from Ludvig et al. 2012) includes the following equations:

(1) Reward prediction:

$$V_t(\mathbf{x}) = \mathbf{w}_t^T \mathbf{x} = \sum_{i=1}^{n} w_t(i)x(i).$$

(2) Prediction error:

$$\delta_t = r_t + \gamma V_t(\mathbf{x}_t) - V_t(\mathbf{x}_{t-1}),$$

where Equation (1) is how a prediction is calculated at a given time, and Equation (2) is how a prediction error δ_t is calculated at a given time.[24] The first term of (2), r_t, is the stimulus intensity (the reward received at t), the second term is the new reward prediction at t (adjusted by discount

---

[24] Equation (1) is (1) and Equation (2) is (3a) in Ludvig et al. (2012). The TD model description also involves equations for modifying the weights $\mathbf{w}$ for the elements of the stimuli and for quantifying the decay of the eligibility trace $\mathbf{e}$ for each weight (the eligibility is a plasticity window for modification of a connection). These complications are not critical for my purposes, but I include these equations here for completeness:

(3) Weights of stimuli ("How much weight should this feature now be given?"):

$$\mathbf{w}_{t+1} = \mathbf{w}_t + a\delta_t\mathbf{e}_t$$

(4) Decay of eligibility ("Which stimulus was most responsible for the reward?"):

$$\mathbf{e}_{t+1} = \gamma\lambda\mathbf{e}_t + \mathbf{x}_t$$

where a = a learning rate parameter, $\mathbf{e}$ = a vector of eligibility traces for each stimulus element, γ = a discount factor, and λ = a decay parameter. In connectionist terms, two connections into a node may be of equal weight but unequal in their eligibility or availability for updating. Equations (1)–(4) are equations A.2–A.5 in Suri and Schultz (2001: Appendix A, 858). While this difference does not matter here, Sutton and Barto's (1981) correlation rule is between input traces (eligibility) and output, not input and output. This reflects their view (1998: 136) that neurons are poorly modeled as simple input–output switches, "with little internal processing power". Note the similarity between the simple views of neurons and bacteria (discussed in Chapter 2). They speculate that the eligibility trace in neurons may be realized by an increase in chemical concentration.

factor γ), and the third term is the result of (1) at the previous time step—it is the prior time step's reward prediction. (2) says, crudely: "I'm off by this much today: here's the reward I got today, plus the reward I predict I'll get tomorrow, minus the reward I predicted yesterday that I'd get today." In (1), the vector $\mathbf{w}(i)$ includes the modifiable weights at $\mathbf{t}$, and the vector $\mathbf{x}(i)$ includes the elements of the stimulus representation. In the simplest case, the stimulus is a single event, rather than a series.

Suri and Schultz (2001) implemented the TD model in an artificial neural network, which computed the prediction signal and adjusted its weights using the prediction error signal. They used the network to model dynamic structure instantiated by real neural populations in prior neurophysiological research. Midbrain dopamine neurons exhibit phasic anticipatory activity: after an animal's learning, activation shifts from firing at an unpredicted reward to firing at a conditioned stimulus. Cortico-striatal (putamen) neurons exhibited tonic anticipatory activity: activity that gradually increased between stimulus and reward (Montague et al. 1996; Schultz et al. 1997; Suri and Schultz 1999: 882; Schultz 2000).[25] The TD-model-trained artificial network's behavior revealed a correspondence between the phasic firing of midbrain dopamine neurons and the prediction error signal, and between the tonic firing of cortico-striatal neurons and the gradual increase of the prediction signal between stimulus onset and onset of expected reward. The simulated prediction error signal was comparable to actual midbrain dopamine neural activity, and the simulated prediction signal was comparable to actual cortico-striatal neural activity. As Suri and Schultz note, this result suggests that phasic anticipatory activity of midbrain dopamine neurons induces long-term adaptation of the tonic anticipatory activity of cortico-striatal neurons.

This use of the TD model in a network simulation showed that real neuronal populations appear to be adaptive elements that learn to predict future rewards in the quantitatively similar sense that humans, monkeys, rats, and other adaptive elements do. This is not observing a neural firing pattern and using that pattern to predict what the subject will do, as in so-called "mind-reading" experiments with fMRI. It is finding structure in neural behavior that is quantitatively analogous to

---

[25] Tonic firing is baseline firing activity that occurs spontaneously. Phasic firing is a transient burst spiking pattern triggered by excitatory synaptic input (Goto et al. 2007).

the structure of reinforcement learning in a behaving animal. The dog is an adaptive system that learns when a reward will appear, and will exhibit anticipatory behavior (salivating) when it has learned. Cortico-striatal neurons, or populations of them, are also adaptive systems that learn when a stimulus will occur, and in consequence exhibit anticipatory behavior (firing). Another relevant similarity is that unreinforced activation in neurons is extinguished just like unreinforced responses in animals (Suri and Schultz 1999: 882). In short, we now have strong empirical evidence that real neurons are adaptive elements in a very rigorous sense.

The standard construal of the TD model includes predicting, antici-pating, and expecting, as well as surprise (discussed below). This construal justifies Suri and Schultz's description of specific firing patterns of neural networks as anticipatory activity just as the dog's salivating is described as anticipatory behavior. Of course, it doesn't feel natural to think of neurons as capable of behavior this complex, let alone to think that psychological predicates such as "anticipates" could possibly be true of neurons. There is certainly no a priori or intuitive reason to group the firing pattern of a neuronal population with the salivation of a dog as anticipatory responses in the absence of a learning model that applies to neurons and dogs. There is no a priori or intuitive reason to think features of classical conditioning in animals that are captured by the TD model would also arise at the cellular level. But what feels natural isn't a good guide to metaphysics. Indeed, Mitchell et al. (2009) suggest that anticipation is ubiquitous throughout biology because it is adaptive.

But how seriously should we take the standard construal in these contexts?[26] For example, in an assessment of the precursor Rescorla–Wagner learning model, Miller et al. (1995: 363) note that "descriptions of the model using the language of 'expectancy' and 'representation' are interpretations; the model itself does not demand that language." As a logical point, this is correct. The construal need not be in these terms. But this is precisely the same as saying that the variables in the Lotka–Volterra

---

[26] Weisberg (2013: 76) notes that modelers' construals are often implicit, and depend on shared background knowledge in a scientific community. That doesn't mean they can't be reasonably inferred or that they are never made explicit (even if they lack the detail one might want in a philosophical analysis). Sutton and Barto are explicit.

equations do not demand predator–prey language in their interpretation. They don't. But the reason the psychological language is used in the one case, and predator–prey language in the other, is that the models were developed to explain phenomena to which these concepts already applied, if for qualitative reasons. Using different concepts would require ignoring the intentions of Sutton and Barto, Rescorla and Wagner, and Lotka and Volterra to develop models that would quantify the dynamics of these qualitatively known relationships. Sutton and Barto (1981: 162) express this intention when they write that the concepts of eligibility and expectation in their model "are not only of critical importance in accounting for animal learning behavior . . . but can be associated quite naturally, albeit speculatively, with certain processing capacities of neurons". Using methods and empirical data not available in 1981, Suri and Schultz (2001) show this is not mere speculation.

3.5 Concluding Remarks

In the last two sections I presented two mathematical models of cognitive processes that have been fit to data in domains as far from the human as fruit flies and neurons. These models are as robust as any that we currently have in psychology. These cases seriously challenge the anthropocentric semantics of psychological predicates because they rely on a well-respected scientific method for discovering similarity that is independent of qualitative similarity to the human case. Combined, the cases discussed in this chapter and in Chapter 2 generate significant semantic pressure on our interpretation of psychological predicates. At the very least, they motivate a response to the advances in science, even if that response is just stricter policing of intuitive conceptual borders.

The process of semantic revision may be illustrated with the psychological concept of surprise. This concept has garnered recent attention because of the predictive coding hypothesis for overall brain function (Clark 2013; see also Rao and Ballard 1999; Friston 2010; Hohwy 2013; Calvo et al. 2016 consider its application to plants). The predictive coding hypothesis holds that neural systems generate predictions from an internal model, compare that prediction to new sensory input, and propagate a signal to the next processing stage if the comparison generates an error signal. This is the conceptual core of the TD model, albeit without its critical reference to time, applied to the brain. Given that "[t]he essence of the TD model is surprise" (Barto et al. 2013), the question has arisen of what "surprise" means in the predictive coding model (although not, to my knowledge, of what "anticipating", "expecting", or "predicting" mean).

A new term, "surprisal", has been coined to distinguish surprise in the model from surprise in a phenomenological, anthropocentric sense. Surprisal is the mismatch between a predicted signal and the actual input, and is distinguished from surprise in the "normal, experientially loaded" sense (Clark 2013: 186). But the concept of surprisal is not semantically discontinuous from the concept of surprise. Both involve violations of expectation, or a discrepancy between an expectation and an observed actuality. For example, in distinguishing the concept of surprise from that of novelty, Barto et al. (2013: 6) note that common-sense and prominent scientific formulations of the surprise concept involve a comparison between an expected and an actual observation: surprise is the result of a process of comparing an input with an existing prediction in which the input and prediction do not match. (Novelty involves comparing an observed actuality with the contents of memory.) We can but need not also have an emotional response to this discrepancy.

This semantic overlap may be one reason surprise and surprisal are not always clearly distinguished (e.g. Friston 2013: 212).[27] So why not call a spade a spade? Surprisal is surprise without the conscious affect commonly associated with but contingent to it. The conscious affect is a stereotypical feature of human surprise that we have traditionally used to characterize a psychological state that in actual fact is not restricted to

---

[27] In contrast, both these psychological uses appear semantically very distant from the term "surprisal" introduced by Tribus (1961) in thermodynamics and used by Shannon (1948) as the basis of the entropy or uncertainty measure in his mathematical theory of communication (entropy is the average of all surprisals weighted by the probability of their occurrence). Clark (2013) notes that "subpersonal" and "personal" surprisal properties can come apart—a percept of an elephant in a seminar room may be least "surprisal-ing" in that it generates the smallest prediction error. But this may just be another way of saying that a person can be surprised (can encounter the unexpected) but not feel stereotypical affect. Meanwhile, Dennett (2013: 209–10) suggests there is a metaphorical relation between "surprisal" and "surprise", in the way we use "projection"—a nominalization of a "familiar and appealing verb"—as "shorthand" to talk "metaphorically" about the literal truth regarding color within the Scientific Image.

humans. We weren't able to figure this out before because we didn't have the scientific tools to do so. In a similar vein, Knutson and Greer (2008) distinguish anticipatory affect—the emotional states people experience while anticipating a weighty outcome—from the anticipating of the outcome. Given this distinction, a neuron can anticipate a stimulus, and exhibit anticipatory behavior, without experiencing anticipatory affect. There is no need to coin a new term ("anticipatal"?) for what it does. Our understanding of anticipating has improved. Just so, we did not coin "goldal" to refer to the element with atomic number 79. We kept "gold" but left pyrite out of its extension. The coining of "surprisal" presupposes a semantic assumption about "surprise" that we have good empirical reason to reject.[28]

Revisions in psychological concepts are being tried out elsewhere as well. Eldar et al. (2016) suggest new ways of conceptualizing mood to understand how it interacts with decision-making and learning, the latter specifically in terms of the TD model. Attention in visual perception is being reconceived by some as "a cortical mechanism for reducing uncertainty" within a broadly Bayesian view of perception as probabilistic inference (Rao 2005: 1847). Berridge (2004) examines the concepts of motivation and reward and distinguishes commonly associated emotional responses from the cognitive processing associated with human emotional responses (see also Wright and Bechtel 2006: 65–72; Roskies 2010). An incentive salience concept of wanting is distinguishable from the hedonic sense of liking (see also Robinson and Berridge 2000). Such adjustments in motivation concepts have been helpful in other theorizing, including theories of schizophrenia (Kapur 2003) and addiction (Robinson and Berridge 2000). In general, neuroscientists are "iteratively finessing cognitive and behavioural concepts" as they investigate neural processes (Seth 2014); from this perspective, "to discard concepts like memory *tout court* from the remit of neuroscience seems rather to throw the baby out with the bathwater" (Seth 2014: 5–6). Ditto for surprise, anticipating, and the rest. In the light of new research we are beginning to

---

[28] I'm not implying that we already know what the emotions are. The same moral goes for them. We may obtain better understanding of the emotions, find out that specific emotions are found in far more living things than we thought, and revise our concepts yet again to reincorporate emotional components properly understood.

develop a more precise, scientifically informed psychological vocabulary for uniform use across human and nonhuman domains.

These semantic adjustments can also reverberate back to the human domain and provide reason to reconsider what the terms mean when applied to humans. Just as a lack of human-stereotypical features need not prevent nonhumans from having the same cognitive abilities we do, we may revise our human psychological stereotypes and their role in what we mean when we use psychological terms for ourselves.

Note that the process of revision does not have to be complete before a term can be useful in science. For example, in distinguishing novelty and surprise, Barto et al. (2013: 2) argue that "[a]lthough the names used to describe results may not be important, the distinction may encourage neuroscientists to ask questions such as: Is there a predictor at play? . . . Or, if there is no prediction, what are the memories that are searched for?" Conceptual clarity is important to philosophers and scientists, but the latter will often work with notions awaiting further precision. Terms used in this way may be considered *usefully vague* (akin to Godfrey-Smith 2005: 4): they have enough sense to mark basic differences, but are not so precise that they conceptually structure a field prematurely.

As a result, one *could* say—although, to be clear, Suri and Schultz (2001) do not—that neural populations exhibiting anticipatory activity *want* the expected reward, just as we say a salivating dog wants the food. Following Robinson and Berridge (2000), "wanting" does not entail ascribing pleasurable sensations ("liking") when a reward arrives on time. So when Dennett (1978b, 1981/1997) famously writes about a colleague who describes a von Neumann computer as wanting to get its queen out early, this psychological ascription can be interpreted in two ways. One is to say that, given what we're learning about wanting, there's absolutely nothing funny going on semantically with this use of "wanting". There may not be any empirical justification for the ascription, but that's a different issue. That is a Literalist type of response. The other is to deny that scientific advances are having a revisionary effect on our understanding of real wanting. True wanting isn't being ascribed to the computer because the term is used with implicit scare quotes indicating a non-Literalist interpretation, such as the Intentional Stance.[29]

---

[29] The Intentional Stance (Dennett 1981/1997) may also be understood in a way similar to the view of many economists regarding preferences (consistent utilities) in classical decision theory—these are "as if" ascriptions that might or might not correspond to any

Of course, the debate is all about whether human emotional responses or other human characteristics are essential to psychological capacities, or alternatively whether the terms or concepts that refer to these capacities are defined in part in terms of these human features. The first description focuses on the metaphysical dimension, the second the semantic dimension.

In the next four chapters, I consider in detail the options for responding to the scientific developments canvassed in this chapter and Chapter 2. Literalism, discussed in the Chapter 4, is a progressive strategy. In the light of the sources of and motivations for the uses, it interprets them as being referentially uniform and literal across human and nonhuman domains. The remaining three responses, discussed in Chapters 5–7, are conservative strategies. They implicitly deny that referential revision is taking place by offering ways of interpreting the new uses that leave untouched the anthropocentric standard for ascriptions of genuine psychological properties or capacities.[30] They do not all deny that the uses should be interpreted literally, but they all implicitly reject the Anti-Exceptionalist metaphysical component of Literalism.

---

internal state as far as their utility (in current theory) is concerned. I discuss the Intentional Stance in greater detail in Chapter 7.

[30] Folk-psychological eliminativism (Churchland 1981) is a kind of conservative strategy: it counsels omitting psychological terms from serious scientific discourse. *Pace* Churchland, psychological terms are not analogous to "phlogiston", which was coined and abandoned within physics; it did not carry any baggage from a previous semantic life. Coinage and extension are distinct ways of introducing theoretical terms in a science; the concern here is extensions. (Strevens 2012 discusses coined or "minted" terms.)

4

Literalism

An Initial Defense

4.1 General Remarks

In Chapters 2 and 3 I presented uses of psychological predicates across four unexpected domains. The uses are selective, systematic, and responsive to scientific evidence of various kinds. Although I did not emphasize this point, all the cases discussed were reported in peer-reviewed publications intended primarily for a readership of peer scientists. These contexts are paradigms of serious scientific discourse. It makes no difference to the plausibility of Literalism what anyone at all might say, even if they say it in *The New York Times*. The relevant kind of question is, what do the research scientists say in *Nature*? The default interpretation of their discourse in these contexts is that it is intended to express truths (or empirically justified claims) directly.

I also did not emphasize that scientists are *not* explaining plant, bacteria, fruit fly, or neuron behavior by populating their heads (or whatever) with beliefs and desires. They are not taking the Intentional Stance towards these entities if this essentially involves belief–desire–intention explanations of their behavior that presume some form of human rationality.[1] Rather, vocabulary follows discovery. We have long distinguished ourselves from other beings by our flexibility and complexity, as measured by our

---

[1] "To a first approximation, the intentional strategy consists of treating the object whose behavior you want to predict as a rational agent with beliefs and desires and other intentional states exhibiting what Brentano and others call intentionality" (Dennett 1981/1997: 59). Ascribing rationality is a second-order normative assessment of first-order cognitive states. We can reason irrationally. In principle, belief and desire are just as subject to scientifically motivated extension and revision as expectations, decisions, and the like, but for the moment they are not being ascribed (systematically or at all) in the relevant contexts. Maybe they never will be.

anthropocentric lights. We have long used psychological predicates to denote mental features associated with and inferred from our flexibility and complexity so conceived. It is no accident that discoveries of flexibility and complexity in nonhumans have prompted scientists to selectively reintroduce components of this conceptual framework into unexpected domains. The successful uses of cognitive models across human and nonhuman domains provide further quantitative evidence of structural commonality for specific capacities. Epistemically, these models provide strong evidence that at least some of the structures we refer to using psychological terms for humans are possessed by nonhumans as well.

So when Dennett (2013: 210) asks, "What is literally going on in the scientific image?", the general scientific situation strongly suggests that we went a bit too far in the transition from the Original to the Manifest Image, and that we are now correcting ourselves in the developing Scientific Image. That, at least, is the Literalist take on it. Literalism holds that psychological predicates are being used with literal intent to pick out the same scientifically-discovered structures across the relevant human and nonhuman domains. The sameness of reference across domains is the metaphysical element of Literalism, which I call Anti-Exceptionalism. There are clear reasons to think the construals of the same mathematical models used across domains are univocal, especially when taking into account the explicit intent on the part of modelers.[2] This implies referential revision relative to the anthropocentric standard for sameness of reference that we've taken for granted up to now. While scientists are not primarily in the business of conceptual development, some are explicitly doing just that to accommodate psychological terms to empirical discoveries.

Consequently, the Literalist holds that the semantic contribution of psychological predicates to truth conditions is the same in statements about the relevant nonhumans as it is in statements about humans.[3]

---

[2] Relatedly, Wilson (2006: 427–9) holds that "water = H2O" enables "some form of the familiar doglegged H-O-H structure" to ground stability of use even if neither term designates a fixed attribute. The identity merely signals a "marriage" of "linguistic fates".

[3] This is a bit imprecise: within pragmatic semantics, truth conditions are very fine-grained, so in those terms the Literalist claim would be that the truth conditions are sufficiently similar to human uses to count as equally literal. Also to a first approximation, reference is the world's contribution to truth conditions and is one aspect of meaning. "Vixen" and "female fox" might mean different things, but they pick out the same property and have the same extension. I discuss these and other semantic details in Chapter 6.

The statements might turn out to be false—maybe bacteria don't communicate linguistically, maybe plants don't make decisions, and so on. The truth of Literalism does not require the truth of the predications, since it is a view about the proper interpretation of the predicates. Clearly I do think the statements are plausibly true (empirically justified), but like any empirical claim they are open to falsification or disconfirmation. This situation is no different from when we claimed "Electrons orbit atomic nuclei", meaning that they orbit atomic nuclei. It turned out the claim was false—electrons do not have classical trajectories—but the literal interpretation of the claim was the correct one.

Whether the conceptual change in this case is called *replacement* of an old concept by a "conceptual descendant" (Nersessian 1992: 10) or *revision*, in which the same old concept is being used in a revised form, is not a substantive issue. This choice of label—replacement or revision—was equally available for "gold" after the discovery of its internal structure and atomic number. The substantive question is whether the scientifically established reference is plausibly understood as capturing most or all our original referential intentions. We agree that it did in the case of "gold", although we should not presuppose that every individual agreed instantly. Literalism claims the same for psychological predicates. The referential revisions to "decision-making" and so on involve making our original referential intentions more precise. Our original referential intentions with many psychological predicates are probably not all that determinate anyway.

It may be that opponents of Literalism are simply not aware of the scientific developments that motivate the Literalist position. In any event, the starting point for opposition is not the Anti-Exceptionalist metaphysics or its empirical justification, but semantics. When the new uses are pointed out, the initial response is almost invariably "They don't mean it literally", where "they" are the scientists and "it" is the ascription of decision-making to plants and fruit flies, anticipating to neurons, and so on. The semantic alternatives to Literalism claim either that the terms don't refer or that their reference differs between human and nonhuman uses. I discuss these alternatives in Chapters 5–7.

In this chapter, I'll elaborate and support Literalism further by means of a familiar argument for other minds. I also clarify what would falsify Literalism. I then engage at length with an imaginary interlocutor, whom I call the Implicit Scare Quoter, to defend the plausibility of Literalism against a series of *prima facie* objections. Those objections that suggest semantic alternatives to Literalism are considered in later chapters.

4.2 Literalism Elaborated

The Literalist response to the scientific developments presented in Chapters 2 and 3 may be non-technically explained by means of a familiar and anthropocentric way we think about the extent of the psychological: an inference to the best explanation (IBE) to other minds.[4] Literalism holds that the scientific uses are univocal unless otherwise indicated (such as by systematic use of scare quotes). The anti-Literalist must motivate a non-question-begging, non-univocal reading as well as articulate and defend her preferred non-Literal interpretation of the uses for nonhumans. My use of IBE is not primarily to argue for Literalism, although it serves that purpose too. It is to show how Literalism might be defended within a familiar framework of debate about the minds of others, and to help expose the way anthropocentric presuppositions play a role in rejecting Literalism.

The IBE argument for other minds (the "Old Argument", based on Pargetter 1984) goes something like this:

(OA)
1.   My behavior is caused by my mental states.
2.   I observe that others behave similarly to me.
3.   Either they have mental states that cause their behavior, or I am unique and something else causes their behavior.
4.   The first hypothesis is best because it explains both cases.
5.   So it is probable (or rational to conclude) that they also have mental states.

As Pargetter (1984: 160) puts it, the other minds hypothesis is justified because of its "explanatory power". It is the inference to the best available explanation of all of the available evidence. Of course, the inference is not

---

[4] Many philosophers consider the argument from analogy weak because it is based on a single case and needs criteria for relevant similarity. But in practice scientists often agree on relevant similarities, single cases can be legitimate bases for inductive inference (Caramazza 1986), and the inference to other minds (if it is an inference) may use both analogy and IBE (Benham 2009). In any case, either argument would do for my purposes.

really just based on others' overt behavior. Our best explanation will also take into account similarity in form and biological makeup, plus familiarity in interaction (Mori 1970). A more realistic version of premise 2 would involve these and other factors, appropriately weighted, but I will leave it as is for the sake of simple presentation.

To a first approximation, we now have the following "New Argument":

(NA)

1.   My behavior is caused by my mental states.

2.   Scientists have developed quantitative cognitive models of my and others' behavior.

3.   Either the models' construals are the same for others or I am unique and we must interpret the construals differently for the others.

4.   The first hypothesis is best because it explains both cases.

5.   So it is probable (or rational to conclude) that they also have mental states.

By "interpret the construals differently" I mean that the terms in the construal are to be assigned distinct reference than what those terms mean when used for me. This is clearer than saying the models have different construals. It is the difference between, say, continuing to use the same word form "predator" in an extension of the Lotka–Volterra model to a new domain while claiming that it now refers to something different, versus substituting a new term in the place of "predator", which makes the intended difference in reference explicit. The fact that the same word forms are used across domains in itself puts a (small) burden of proof on the anti-Literalist.

(NA) is a way of putting the other minds' argument in modeling terms. To keep a strict parallel with (OA), one must assume in 2a that I am the scientist and that I developed the model for myself.[5] Now, however, the explanatory power of the inference to other minds is based on the formal

---

[5] Of course, scientists make this inference to my mental state too. In this post-behaviorist era, manifested and measurable behaviors (e.g. response time differences) are empirically necessary to justify an inference to a posited cognitive process. Researchers design experiments to elicit behaviors that will license such inferences or help adjudicate between competitor models of the cognitive processes to which they may infer. If the behavior predicted by a model is observed, the inference to its posited cognitive states is *prima facie* justified and the model is confirmed to a degree (or has survived an attempt at falsification).

similarity established by the model. I may notice that you are like me in many ways, but my inference rests on a quantitative model that (per rather ludicrous assumption) I developed on the basis of my behavior. I continue to rate as better an explanation that uses fewer resources to explain more. That's why the first hypothesis in each argument is rated better.

To elaborate (NA) using the DDM discussed in Chapter 3, when I am asked to decide quickly whether a picture is of a house or a face, my response times lengthen and my accuracy drops when I am presented with blurry pictures. From my perspective, I know that my response is an effect of my decision-making processes (my "mental state"). The key to the inference is whether I should construe the DDM as being about decision-making processing in others just as I do for myself. For example, it is whether the label "decision time" that I use in my construal of the equations refers to the same process in others as it does in me. We can make this explicit in a special-case version of (NA):

(NA-SC)
1.   My choosing behavior is caused by my decision-making processes.
2.   Scientists have developed a quantitative model of decision-making of my and others' choosing behavior.
3.   Either the models' construals are the same for others or I am unique and we must interpret the construals differently for the others.
4.   The first hypothesis is best because it explains both (or all) cases.
5.   So it is probable (or rational to conclude) that they also have decision-making processes.

This inference is still perfectly cogent to anyone who accepts (OA) or its reformulation as (NA). Although (by assumption) I developed the model for myself, I find that it applies to others as well. But here's the thing. It just so happens that the *others* include fruit flies, *inter* who knows how many *alia*. Who knew? (Volterra started with fish, after all.) What I as a layperson thought fruit flies were capable of, before testing my model on them, is irrelevant. I don't care how similar they are to me by my untutored observation. I've got a hypothesis, and a model, with more explanatory power than I dreamed of. My prejudices should not get in the way if I want the best explanation of all the cases.

In this way, we can see that the argument to other minds is usually understood in a way that inserts an implicit domain restriction of "others" to humans and anything sufficiently qualitatively similar to humans. This inference is based on lay observation and maybe a little scientific knowledge—*but not too much*, because too much science threatens the implicit domain restriction. What we as laypeople, even moderately scientifically literate laypeople, think about fruit flies is limited and no doubt biased. It is evidence, but no longer our best evidence. How we think the predicates should be understood in the light of this epistemic change is the issue. The Literalist says we should not let our entrenched biases affect our understanding.

(NA-SC) raises this problem in terms of a single cognitive model at an early stage in cognitive research. But it is surrounded by a wealth of new biological knowledge that is increasingly conceptualized in psychological terms because it promises to promote understanding, generate testable predictions, and suggest fruitful new research. In other words, Literalism is not exactly out in front on the issue. The widespread use of psychological predicates across biology shows that many scientists have *de facto* abandoned the epistemic priority of lay observation in their inferences to others' cognitive states. Phenomenological similarity to humans and human behavior constrains new scientific *usage* not at all, even if this usage leaves a semantic puzzle in its wake. When a norm no longer guides a practice, it has been abandoned by practitioners. One need not assert P is false to reject P; one can just go on with one's life, or one's research, unconstrained by a commitment to P. In this case, what is disregarded is the idea that there is a conceptual barrier to the new uses. Those in science who question the new uses are concerned about their scientific utility. Even if there *really is* some such conceptual barrier and scientists are rampantly violating it—as the Nonsense view (Chapter 5) holds—it does not *in fact* constrain science, whether or not one thinks it *should*.

Literalism does not claim that all the psychological properties we possess are possessed by every other biological entity. But the phrase "the psychological properties we possess" bears scrutiny to clarify exactly what Literalism rules out.

(1) *The psychological properties we possess* are *properties that are exclusively human.* This is compatible with Literalism if it means no more than saying that (e.g.) the human visual system is exclusively human in the same way the human hand is exclusively human.

Human-typical psychological properties reflect our "species-specific window of viability" (Clark 2013: 193) in relation to the world. The mantis shrimp's visual system is *exclusively mantis-shrimpy*, too.

(2) *The psychological properties we possess* are *the psychological properties that there are (in the actual world)*. This is also compatible with Literalism. It could be that all the discoveries and models I have highlighted are wrong, that this is all bad science, and that as a matter of empirical fact humans are the only possessors of psychological properties. In this case, the new uses of psychological terms are false. But Literalism does not claim that the statements must be true, only that they are best interpreted as literal with sameness of reference across human and nonhuman domains. For example, Trewavas (2014) claims that plants exhibit a form of consciousness that is distributed throughout the plant rather than localized, as it is in our brains. The Literalist says the term "consciousness" makes the same contribution to the truth conditions of his sentence as it does when it is used in a sentence ascribing a form of consciousness to humans (whatever contribution, exactly, that is). It does not hold that his claim must be true.[6]

(3) *The psychological properties we possess* are *the psychological properties we phenomenologically take ourselves to have*. That we have a specific perspective on our psychological properties is also compatible with Literalism. It does not follow that what we grasp phenomenologically is correct or is the whole story or even the most important part of the whole story. Once upon a time, around 600 BC, water did not have a hidden essence. Some—Thales, at least—thought it was the hidden essence of everything else. Much later, our concept of water—that stuff—adjusted to scientific theory and discovery in a way that clarified our original referential intentions. Our psychological concepts are open to similar adjustment even if we begin from a phenomenological standpoint. Our phenomenological appearances may stay the same, but this does not entail taking our appearances as the last

---

[6] Zink and He (2015: 724) remark that "most philosophers would be skeptical" of Trewavas' claim. But there are non-skeptics among philosophers (Chalmers 1995) as well as neuroscientists (Koch 2012, for a popular audience).

word about their reference. Mathematical models provide an objective basis for understanding similarity of states that once upon a time we could *only* grasp phenomenologically from the first person.[7]

(4) *The psychological properties we possess* are *the psychological properties we phenomenologically take them to be and this is what they really are.* This claim is incompatible with Literalism, since it entails that we already know what psychological properties really are and that this knowledge is based entirely on observation of ourselves. This claim constrains the proper extension of psychological predicates to features of human embodiment, behavioral manifestations, experiential accompaniments, processing bottlenecks, social aspects, and so on that we are familiar with (or just learning about) in our own case. It claims that we do not just understand mental states *in* our own case; we understand mental states *from* our own case—where "our" case is the human case, not me. (How willingly we ignore the problem of other minds when it suits us!) That is the upshot of the implicit and undefended domain restriction of "others" in the above arguments to other humans. Literalism claims this domain restriction is scientifically unjustified. We're only starting to find out what the proper domain should be, with the help of new tools that enable us to escape the parochialism of our phenomenal perspective. Cognitive states and processes are potentially present in lots of beings however different we intuitively think them to be. Literalism does not claim that every psychological predicate will end up with a scientifically determined reference, just as orichalcum did not get its own square in the periodic table of the elements. The Scientific Image can embrace some concepts or terms and ignore others.

---

[7] Along similar lines, Rakova (2003) defends a kind of Literalism for double-function and synesthetic adjectives, such as "hot" and "bright". Double-function adjectives modify physical and psychological states (for example, "bright light", "bright student"). On her view, these terms express a single amodal concept caused by a single property realized in different sense modalities. We share the view that these sorts of uses are equally literal across domains and that science will inform us about the unifying properties. However, Rakova holds that evidence of a unitary property is provided by common neural activation (e.g. the same brain areas are activated for uses of "bright" in "bright light" and "bright student"), rather than formal models.

Literalism does not deny that humans are special or that our psychological capacities are special. In fact, it is deeply non-reductive. More precisely, it suggests that the old debate between reductive vs. non-reductive physicalism may not make sense in a scientific context in which the same formal structures and construals can be explanatory at multiple scales (I return to this point in Chapter 8). The complexity of our own case and its importance to ourselves is also not in dispute. But what makes human psychological capacities special need not also be why they are psychological. We can be unique by using the cognitive tools we possess in species-specific ways in species-specific contexts. Literalism also does not impose any a priori restriction on where psychological language may be properly used, recognizing that we are unlikely to discover a bright line between the haves and the have-nots. We can't even do that for life. The current restriction to biology is empirical, not conceptual or logical. It is almost certainly temporary given advances in artificial intelligence, although for present purposes this doesn't matter.

4.3 The Implicit Scare Quoter

Literalism may be considered the cogent conclusion of the IBE argument to other minds, updated to reflect the new form of evidence and shift in evidential priority that are restructuring our understanding of psychological properties and concepts. The intuitive plausibility of Literalism is another matter.

The easiest way to present the intuitive objections is by using an imaginary composite interlocutor: the Implicit Scare Quoter, or ISQ. The ISQ articulates a number of objections that have been made frequently in conversation with me by scientists (mainly neuroscientists and psychologists) and philosophers. The ISQ is so-called because the objections often start from the claim that when a psychological predicate is used in these nonhuman domains it is surrounded by implicit scare quotes. The debate begins not with another semantic interpretation or a different metaphysical view but with the simple denial of a literal interpretation. Not incidentally, the usual lack of explicit scare quotes or other conventional signs of meaning change in the peer-reviewed literature must be explained away by non-Literalists. As with the fact that the same word forms are used across contexts, the lack of systematic uses of scare quotes indicates a burden of proof on the non-Literalist side.

My goal in this section is to show that these initial objections do not stop Literalism in its tracks. I will focus the discussion by using the term "prefers", given its systematic and longstanding use for neurons and humans, *inter alia*.[8] Here are some examples:

(1) In *preferring* a slit specific in width and orientation this cell [with a complex receptive field] resembled certain cells with simple fields. (Hubel and Wiesel 1962: 115)

(2) It is the response properties of the last class of units [of cells recorded via electrodes implanted in a rat's dorsal hippocampus] which has led us to postulate that the rat's hippocampus functions as a spatial map.... These 8 units then appear to have *preferred* spatial orientations. (O'Keefe and Dostrovsky 1971: 172)

(3) A resonator neuron *prefers* inputs having certain frequencies that resonate with the frequency of subthreshold oscillations of the neuron. (Izhikevich 2007: 3)

(4) Capuchins...*prefer* ready-to-eat fruit but when pressed are capable of pounding apart hard nuts, stripping tree bark, raiding beehives, and even killing small vertebrates. (Chen et al. 2006: 523)

(5) [S]ix-month-old infants can discriminate attractive from unattractive faces and they visually *prefer* attractive faces of diverse types. (Langlois et al. 1991: 82)

For the Literalist these are all literal ascriptions, all on a par referentially.

I deliberately engage with the ISQ over the example of "prefers" used for neurons for three reasons. First, ascribing preferences to neurons is a lightning rod for intuitive objections to a Literalist interpretation. Ascribing decision-making to fruit flies doesn't even come close as an effective (and affective) intuition pump. Second, neuron preferences are a hard case for the Literalist, because the uses are based on qualitative analogy. A mathematical model of preferring is not available, although one might well develop out of ongoing work on motivation and reward. These uses do not have the strongest kind of evidential backing, even though the evidence on which the ascriptions depend requires impressive

---

[8] For what it's worth, a Google Scholar search (in fall 2015) for "neuron believes" (exact phrase) comes up with 1 result (Deneve 2008: 98), where "believes" is in explicit scare quotes. A search for "neuron prefers" (exact phrase) gets 133 results, and no scare quotes. (A reference to Colby (1991) as putting "prefers" in explicit scare quotes is incorrect.) In any event, uses of scare quotes in the scientific literature are highly unsystematic.

technical machinery and neuroscientific knowledge and expertise. In effect, I am voluntarily tying my dominant hand behind my back. Third, because neuroscientists may also feel the force of these intuitive objections, these uses of "prefers" threaten to make Literalism anathema to the very scientific communities whose linguistic practices it seeks to show are systematically rational and justified. This book is not *The Secret Life of Neurons*. Literalism does not claim or imply that neuroscientists who ascribe preferences to neurons think neurons are little people. But the intuitive objections show how deeply anthropocentrism informs our intuitions about the meanings of psychological predicates.

In addition, it might turn out that individual neurons don't prefer but only neural assemblies do, or that neurons or neural assemblies don't prefer at all, once the reference of "prefers" has stabilized. By the same token, maybe capuchins and human infants won't prefer either when it is stabilized, and contemporary cognitive ethologists and developmental psychologists are wrong to think they do. Literalism is committed to the claim that all the sentences above count as fact-stating discourse and that the semantic contribution of "prefers" is the same in all of them. For the sake of argument, however, I am assuming that they all do assert truths.

Finally, Literalism is a claim about the best overall interpretation of all the new uses, even if I focus here on the case of "prefers". The uses it targets are systematic, specific in the types of capacities ascribed, driven by empirical research, and selective in the types of entities to which they are ascribed. This means the non-Literalist must do more than just provide an alternative interpretation of an isolated use of one term in one domain or even one instance. The non-Literalist has to defend a kind of error theory whereby, despite all appearances and evidence to the contrary, the unexpected uses are not what they seem and her alternative interpretation is principled and not ad hoc. That is, her rejection of this specific case must be grounded in a non-question-begging principle that semantically distinguishes human and nonhuman uses of psychological predicates, since Literalism entails that there isn't one. Otherwise her rejection of a Literalist interpretation of "prefers" for neurons is just a rejection of a Literalist interpretation of "prefers" for neurons. The ISQ does not have to defend her principle here, but she'll need to do so when she follows up her rejection of this case with a positive account of her view.

After parrying Literalism with an initial claim that the terms are surrounded by implicit scare quotes ("Not-Literalism!"), the ISQ typically proceeds as follows:

(#1)   The scientists use *"prefers"* in neuron sentences as a technical term with a different meaning; that's what the implicit scare quotes signify.

(#1) is just another way of relabeling the ISQ's initial rejection of Literalism. All she has done is to restate her initial claim that the term differs semantically from the human cases. For example:

(6) Infants *habituated* to two jumps of a puppet *dishabituated* when shown three jumps, and vice versa. (Wagner and Carey 2003: 165, citing Wynn 1996)

(7) The time course of *habituation* of the responses of another neuron (be0282) that *habituated* in one trial are shown in Fig. 5. (Rolls et al. 2005: 116)

"Habituation" is a technical cognitive term (as is "dishabituation") that refers to a non-associative form of learning (Rankin 2004: R617 Box 1). A technical cognitive term in scientific uses can still be used with the same reference for humans and neurons. The fact that a term is technical tells us nothing about its semantics. In fact, technical terms often retain important semantic links to their non-technical cousins. For example:

But the earnest desire to look on blood and death, is not peculiar to those dark ages; though in the gladiatorial exercise of single combat and general tourney, they were habituated to the bloody spectacle of brave men falling by each other's hands.   (Scott 1820/2008: 479)

These links often help account for their utility in new contexts, as we saw in Chapter 3 with "surprisal". Virgin coinage of the "quark" variety is rare. These links also make technical terms vulnerable to being misunderstood by laypeople. But this fact hardly tells us in what the misunderstanding consists. Is what's left out human-specific semantic baggage, or is it essential to its reference? Without being told in what the difference consists we have not been given a reason to think Literalism is false. The same technical term could be used for (1)–(5), as appears to be the case in (6) and (7) for "habituate", and it could have the same reference as in lay contexts. Creating a technical term by shaving off aspects of the meaning of an old term or its connotations can leave behind its actual reference, at long last revealed without the usual, but contingent, human associations.

The ISQ takes up the challenge:

(#2)   In neuron sentences *"prefers"* refers only to response selectivity. We could just substitute *"response selectivity"* for *"preference"* in neuron sentences. For example, a neuron's preferring to fire at horizontal lines is just its responding selectively to horizontal lines. We could just point to the highest point on a graph of a neuron's tuning curve, which plots firing rate as a function of stimulus type.

But we could substitute *"response selectivity"* for *"preference"* in (4) and (5), and we could plot reaching behavior or looking time as a function of stimulus type. In fact, some of us *do* plot infant reaching behavior and looking time as a function of stimulus type, and those graphs correspond to neural tuning curves in terms of exhibiting preferences.

Of course, in ordinary life we do not track people quite so systematically or use statistics-calculating software like SPSS to reveal a statistically significant relationship between stimulus and behavior. But we have made progress. For now the question is: what is *"prefers"* supposed to involve that *"response selectivity"* lacks? Neuroscientists above all are well aware that neurons are not inanimate, deterministic, input–output devices—just in case that is what the ISQ intends to insinuate by means of the substitution. So let's make what may be insinuated explicit: what is the difference?

The ISQ continues:

(#3)   *Preferring* in humans is a complex internal state that causes these patterns of behavior, whereas the analogous patterns of behavior exhaust the reference of *"prefers"* for neurons. Neurons lack intentions and other mental states involved in human preference. They just respond to the appropriate stimulus. In contrast, we cannot predict what a human will do unless we know what they believe, desire, and so on. We can even respond selectively to what we *don't* prefer.

There is no question that neuroscientists learn not to build into their uses of "prefers" everything that an uninitiated layperson might associate with preferring when the capacity is ascribed to humans. But the issue is whether what is left out is essential to preferring, as opposed to a familiar and anthropocentric accompaniment of it. A history of restricting psychological capacities to ourselves does not prove we are right to do so.

First, what features associated with human preferring are actually essential to preferring, even intuitively? If I am told (or observe) that you prefer Coors to Yuengling, I do not have to know anything else to reliably predict which beer you will select when we next go to a bar. Some infants prefer nursing from one breast rather than the other, or perhaps they prefer lying on one side rather than the other: two choices, measurable bias, very predictable. Many adults prefer sleeping on one side: few choices, measurable bias, very predictable. Genuine human preferences are often ascribed, literally, without invoking additional mental machinery. The additional machinery is often invoked instead to explain exceptions to the reliable patterns of behavior we typically manifest. In rational choice theory, the stability of our preferences, as exhibited in and inferred from our choice behavior (that IBE argument again!) is presupposed. Where there is no such stability, there is no human preference either—at least not unless the ISQ is committed to dispositions that are never manifested or are always perversely or irrationally manifested.

Second, it's true that in this initial response I have focused on one of the lexicalized (conventionally semantically encapsulated) senses of "*prefer*": the sense involving consistent choice behavior rather than the sense involving an affective response. And ISQ #3 rightly notes that humans can consistently choose what we don't like (and too frequently must). At other times, we like what we have chosen (or what has chosen us: Aronson and Mills 1959), and our preferences are influenced by those around us (Izuma and Adolphs 2013). In short, our liking and our choosing can come apart, and their relation is often context-dependent.

But we also don't think that the preferences we exhibit without conscious affect are not real preferences. Many ascriptions of preferences to healthy adult humans, human infants, impaired human adults, and human utility maximizers don't require liking to be real. Preference may essentially require assignment of relative value, but the process need not be understood in a way that depends essentially on specifically human conscious experiences and interpretations of them. Of course, we could develop a new term that encapsulates all the features of the most sophisticated human cases of preferring, in the same way that "*pirouetting*" lexicalizes the features of the most sophisticated human cases of rotating. Philosophers in particular often build additional elements into folk psychological concepts (Godfrey-Smith 2005: 13–14; Glock 2009). But "prefer$_{sophisticated}$" would not be the same term with which the debate with the ISQ began.[9] It would be a new technical term applicable only to specific humans at specific times, at best.

Third, the claim that neurons "just" respond to the appropriate stimulus insinuates the simplistic input–output view of neurons that Sutton and Barto (1981) deplored long ago. But while neurons can be treated methodologically as black boxes in science—the way economists treat utility maximizers, a.k.a. people—they are eukaryotic cells operating in complex networks and living and responding differentially in and to complex environments. Their responses are not 100 percent predictable, as Hubel and Wiesel's (1962) results show. If laypersons are not interested in explaining their unexpected responses, it does not follow that their internal machinery plays no role, whether scientists already know how it does or not.

However, the fundamental issue raised by ISQ #3 is not the mere possession of internal machinery relevant to preferring, but the conceptual nature of that machinery. The ISQ points out that we explain unpredicted human cases using beliefs, desires, intentions, and other intuitively sanctioned cognitive tools (that time you chose Yuengling over Coors: were you trying to impress someone?). She then argues that there is simply no need to invoke these tools to explain neuronal behavior—and presumably not that of many animals, for that matter.

The problem is that the ultimate explanatory power and role of these intuitive tools cannot be taken for granted, even for humans. The models introduced in Chapter 3 are among our first and to date foremost attempts to refine or replace at least some of the items in this intuitive

---

[9] Gunderson (1969: 418 fn. 15) makes a similar point—just substitute "ISQ" for "vitalist" and "preferring" for "self-adaptiveness":

> It is logically open to the vitalist to revise his criteria in such a way that they single out special features, if any, of human self-adaptiveness which are not satisfied by even self-adaptive machines. *However, even to admit the need for such a tightening up of criteria constitutes some substantive retreat.* (his italics)

Rational agents were once thought to require self-reflective consciousness and propositional attitudes (Cummins et al. 2004). See also Healy (unpublished), who argues that theoretical progress in sociology is often hampered by insistence on nuance. The point can be extended, however selectively, to philosophy and other fields.

toolbox with better tools. Ultimately, we would like to be able to explain intuitively simple and complex cases with a few top-quality basic tools—motivation, memory, decision, prediction, affect, and so on—that are constrained in different ways in different contexts to obtain the relevant range of behavioral results. Since we do not know what we will end up with, it is at the very least not clear how we will best explain exceptions to preferences, or any other trait-like state, in humans.[10]

It is also becoming increasingly clear that what's in the human head explains far less of our behavior than we intuitively think. The ISQ presupposes that the explanantia of human behavior are wholly or primarily internal and mental, when we have good empirical reasons to think that the internal machinery (however conceptualized) is far from the whole of the explanation and often not even the most important factor in any given context. Violations of expectations of others' behavior, encapsulated in trait ascriptions, are compatible with basic internal machinery operating in complex, changing contexts, such that the burden of explaining deviation lies in what is outside the head.

In sum, we have not yet been given a reason to think that the term "prefers" does not express a concept that is properly ascribed to humans and neurons. We do not know what human-specific features intuitively associated with preferring are constitutive of it.

The ISQ plows on, not quite sure what to make of this unexpected resistance:

(#4)   But it must be at least *possible* for preferences to interact with other mental states for them to count as preferences. That's what makes the difference between neuron and human preferences a difference in kind, not degree.

---

[10] Gunderson (1969: 407–8) also addresses a similar issue, comparing humans and computers:

> [I]t is possible to know that a given predicate phrase such as 'solves problems' applies to two different subjects . . . and fully understand the conditions underlying the predicate's applicability in the one case without fully understanding the underlying conditions in the other case. And this may take place without there being good reason to suppose that 'solves problems' is ambiguous or has different senses in the two cases.

His point also holds when we do not fully understand the conditions underlying the applicability of the predicates to humans. Linguists draw a distinction between ambiguity and imprecision: an understanding of a term may not be fully specified in a sentence, but it does not follow that the term is ambiguous (Zwicky and Sadock 1975: 4).

Given the answer to ISQ #3, it is not clear what justifies this condition on real preferences. I can agree that neurons do not possess the complexity, cognitive and otherwise, of humans; that is not empirically plausible. But the question was, and is, whether all this complexity, whatever it might turn out to be, is essential for real preferring. And if the objection is a veiled way of saying that psychological capacities must be holistically possessed, then it is simply an assertion that may well turn out to be empirically false. We know that specific cognitive capacities can be absent without general loss of cognition. Patient H.M. is a clear example: his severe loss of the ability to create new memories did not impair his general intellectual capacities (Squire 2009). To put the point in terms of propositional attitudes: the propositions we can entertain may or may not be holistically grasped, but the attitudes towards them are not.

At this point the ISQ starts to get defensive:

(#5)   If none of this complexity matters, you have implicitly deflated the concept so that *preferring* is just a minimal capacity that humans and neurons share. The models that might involve this kind of preferring will be trivial. I *give* you this profoundly boring type of *preferring*. What we really care about is the non-boring variety, found in humans but not neurons.

It's not clear why we should think a feature transforms from being interesting when it is exclusively ours to being banal when it isn't. But the basic problem is that the objection simply sidesteps the issue of conceptual revision. Cross-domain uses of a term can lead to the creation of either very abstract cross-domain concepts or very domain-specific (or subdomain-specific) concepts, keeping in mind that the useful categories we develop as empirical knowledge accumulates are not decidable a priori. But "*preferring*" has not been deflated just in virtue of its being used in nonhuman contexts. A deflationary interpretation is certainly a possible non-Literalist alternative interpretation of the uses, and the ISQ is within her rights to suggest this alternative. But what she has not done is provide a reason to think Literalism is false, and that is what we are looking for now.

The objection also raises problems for the ISQ. Any use for nonhumans is bound to differ in some way from human uses. The ISQ must articulate the principle under which she declares a use to involve a different (deflated) word. For example, if capuchin monkeys prefer grapes to oat cereal (Fontenot et al. 2007), the ISQ must say that either this use of *prefers* is a new deflated term or not. What principle guides this a priori division of senses, and how long is the ISQ's list of senses (Godfrey-Smith 2005: 12)? Is the claim that every psychological predicate used in the new domains is deflated? If not, what makes her deflationary interpretation non-ad hoc? This is the problem of determining the boundaries of model families (Chapter 3) all over again. It is also the problem of providing a principled reason to reject Literalism. This is not done by tossing out a possible alternative semantic interpretation of "prefers" or other specific predicates for individual cases where she just can't stomach a Literal interpretation.

The ISQ now flourishes what she thinks is her trump card:

(#6)   But we live in a space of reasons, and neurons don't. Our behavior is guided by (responsive to) norms, expressed and enforced in practices of giving and asking for reasons. Preferring and other psychological concepts characterize the causes of our actions in terms that display our rationality and make what we do normatively intelligible (responsive to reasons). Neural activity is not guided by norms. So they do not have real preferences, and neuroscientists aren't using the same concept of *preferring* for them.

It is by no means clear how big this space of reasons is or its relation to our behavior, as noted above (e.g. Haidt 2001).[11] It is also false that lacking a particular kind of cognitive state entails lacking them all. Neurons could have real preferences but not have real hopes. However, for the sake of argument I will set aside these issues in order to flesh out the objection sufficiently to enable a Literalist response. I draw mainly on Brandom's (1994, 1997) elaboration of Sellars' (1956/1991) logical space of reasons, although others have provided alternative accounts (e.g. McDowell 1994).

Brandom (1994) provides an account (in the first instance) of linguistic meaning as use, where use is elaborated specifically in terms of social or communal practices of "deontic scorekeeping": practices that express attitudes of commitment and entitlement—that is, the giving of and asking for reasons. The states of those engaging in these practices acquire

---

[11] As Dennett (2014: 58) also notes: "Much of the activity in the space of reasons involves misleading feints and delusional justifications."

their propositional or conceptual contents by playing suitable roles in the reasoning that generates the practices. This is a version of inferential or conceptual role semantics, where the concepts depend on the social uses, rather than the uses presupposing a conceptual system. Note that the nodes in the space of reasons can be occupied by (states of) the beings that trade in reasons (e.g. humans in social networks) or else by the reasons they trade in (e.g. propositions). Depending on this choice, the objection can go by two routes. Neurons do not have real preferences because either (a) their states are not states of entities to which we offer reasons or of which we ask for them, or (b) the states neurons are in do not have contents that, in principle, can be used as reasons or for which reasons can be asked, and so their attitudes towards these contents are not real attitudes. Since *real* preferences are attitudes towards *conceptual* contents, since what makes content conceptual is that it used in a way guided by social practices of giving and asking for reasons, and since neural contents are not used in this way or neurons are not in our social networks (or both), neurons do not have real preferences.

While it is true that neurons are not in *our* social networks and that the contents of neural states do not figure in *our* scorekeeping regime, we still need to know why these are conditions on *real* attitudes.[12] It is compatible with Literalism that real attitudes depend on social networks, as opposed to being individuated, *qua* attitude-type, purely individual-istically. Brandom's framework does not rule out *nonhuman* cognition; it rules out cognition in the absence of the appropriate social or communal structures. Neurons are paradigmatically in networks. As an objection to Literalism, what we need is an argument for why the aspects of human social relations that are human-specific also determine what's cognitively real. We need a non-question-begging defense of the assumption that our specific way of manifesting cognition, socially as the case may be, determines or reflects the metaphysical facts. It might instead be a special but non-standard-setting case of a general phenomenon that is mani-fested with equal reality elsewhere. Why think the real and the human are co-extensive? The Manifest Image encourages and enshrines this assimilation, but that doesn't make it correct.

---

[12] Hurley (2003) and Glock (2009, 2010) argue that non-conceptual contents can be reasons—the space of reasons is not restricted to conceptual contents. That's fine by me, but since I reject the anthropocentric standard my response differs from theirs.

An example from Brandom (1995: 896–7) can illustrate the point. In his terms, what distinguishes the non-conceptual activity of a parrot that squawks "That's red!" from the conceptual activity of a human who says "That's red!" is that parrots do not have or understand the concept of red; the squawk is not the expression of a conceptual content.[13] Both human and parrot can share "reliable differential responsive dispositions"—a.k.a. response selectivity—but only humans commit themselves to and are held accountable for what they express themselves as having a reliably biased differential disposition towards (in this case, red objects):

> Non- or prelinguistic animals do not have status or standing in the space of reasons. So . . . they neither deploy concepts, acquire beliefs, nor count as having knowledge. Nonetheless, it is common to talk about them loosely as though they were capable of some version (usually admitted to be degenerate cases) of these accomplishments. The informational states most closely resembling genuine beliefs that they *do* have (call them *beliefs**), when they both correctly represent how things are and are acquired by a suitable reliable process may be called knowledge*. . . . This state has in common with the genuine article what the parrot has in common with the reporter of red things: reliable differential responsive dispositions.   (Brandom 1995: 899–900 fn. 3)

From a Literalist perspective, the error is to downgrade the parrot's cognitive capacities to belief*—*mutatis mutandis* with neurons and prefer*—because it does not have standing in a human social network (the (a)-route) or because it cannot entertain human concepts (the (b)-route), or both. If what counts as linguistic is a pragmatic issue, then nonhuman communication systems can be linguistic given the right pragmatics. The parrot's "That's red!" could (in principle) express a belief in its own communication system with its conspecifics, and it could very well understand what this string means in its system, and the system can be guided by norms appropriate to it. (If the response is that it merely behaves "as if" it understands its own calls, the dispute

---

[13] Frequently, this is expressed imprecisely by saying its squawk is not the expression of a belief. The imprecision can mislead if the claim is understood in a way that builds into the concept of belief (the attitude or state-type, not the proposition) the idea that real beliefs have human conceptual content. This is just another version of the assumption being scrutinized in the text. By analogy, if we held that a real visual state is a state that has human visual content, one could say with equal imprecision that the parrot doesn't have visual states. The Literalist, alert to the ambiguity, would take this to express "The parrot doesn't have *human* visual states"—true, but no argument against Literalism.

shifts to real understanding.) Brandom's "degeneracy" charge presupposes an anthropocentric standard for any space of reasons; the "looseness" (non-literalness) of the cognitive ascriptions follows. But this is
exactly the question that ISQ #6 cannot beg.

Note that it is orthogonal to this discussion whether the normative
can be reduced to the natural. It is common to contrast the normative
and the natural—for example, by contrasting the space of reasons as a
network constituted by normative relations with the space of nature. Of
course, labeling the dichotomy "normative vs. natural" is misleading if
normativity just is natural (specifically, based in evolution: Millikan
1984; Dennett 2013). But a non-naturalist Literalist (an odd but logically
possible duck) could hold that all psychological properties are normative
in some non-evolutionary sense, *and* that this kind of normativity is
found throughout nature. Everything has a divine spark. The relevant
issue is whether the way in which human social structures are organized,
policed, and maintained over time yields a variety of normativity not
found in nonhuman networks *and* that this specific kind of normativity
is what determines real cognition rather than just the human variety of
real cognition.

In sum, "the" space of reasons is usually tacitly understood to be the
space of *human* reasons.[14] This is no more harmful in principle to
Literalism than our tacit understanding that discussions of "the" brain
are usually about the human brain unless otherwise specified. The
Literalist denies only the imperialistic pretensions of the human case.
From her perspective, it is entirely superfluous to the pragmatic infer-
entialist theory and its validity to use it to establish and police the borders
of the psychological.[15] Even if it is true that genuine intentionality
requires socially-based normative assessment, the account does not
rule out *nonhuman* thought; it rules out *nonsocial* (or *noncommunal*)
thought. But human social relations, networks, and environments can

---

[14] Actually, not always tacit: Brandom's (1994: 4) answer to the question "What is it that
we do that is so special?" is the traditional view that "we are distinguished by capacities that
are broadly cognitive". But this can be true compatibly with Literalism if what is special to
us are human expressions of cognitive capacities.

[15] Pendlebury (1998: 144–5) makes a similar point. Dennett (2010) agrees with Brandom
about the necessity of the social or communal for the intentional, but argues that the social or
communal must in the end have an evolutionary explanation. So even if the relevant relations
are "normative", that doesn't make them the sort of thing nonhumans can't stand in.

make human cognition what it is without making cognition what it is. In the near future, network models applied to social relations are very likely to play the same revisionary role with respect to deontic scorekeeping relations that cognitive models are already playing with respect to decision-making and the like. Network modelers have already trained their sights on neural and human social networks alike (e.g. Aral and Walker 2012; Baronchelli et al. 2013; Bentley et al. 2014). It is very much an open question what specific formal network models may apply to both and—reiterating the point made in Chapter 3—the extent to which social concepts that pick out human relationships may be extended to nonhuman relationships (Figdor forthcoming).

Neisser (1967: 306) noted long ago that "a really satisfactory theory of the higher mental processes can only come into being when we also have theories of motivation, personality, and social interaction. The study of cognition is only one fraction of psychology, and it cannot stand alone." The fundamental issue raised in ISQ #6 is the status of human relationships, traditionally understood, as the standard for the relevant relationships. It is the issue of the anthropocentric semantics of psychology conceived socially and pragmatically.

At this point the ISQ (rather desperately) tries a new, empirically-minded, tack:

(ISQ #7)   Why not just ask the neuroscientists what they mean by "prefers"?! Isn't this just an empirical issue, and one that *they* get to decide? Surely the vast majority of them would emphatically deny that their attributions of preferences to neurons should be taken literally. It would be strange if the *philosopher* could answer the question in one way or another despite the scientist's protestations. The authority seems to belong to the neuroscientist, not the philosopher. *Ha!*[16]

The crux of my response is this: The neuroscientist is certainly in charge of her intended reference of "prefers" in neuroscience, but it hardly follows that her intended reference amounts to a rejection of Literalism.

---

[16] My thanks to two anonymous reviewers for OUP for pressing this objection. In what follows I'll set aside a technical issue in philosophy of language (but see Chapter 6): while the objection is stated in terms of a poll about meaning, fundamentally it is about reference—what in the world a term picks out. The Literalist can agree that the *meaning* of "prefers" differs between these contexts, given the mainstream view that meaning isn't reference (and for some is highly context-dependent).

What the neuroscientist cares about is the utility of the term in her theories and explanations. This goal is satisfied even when she leaves her view of its reference open to revision in the light of additional evidence. She may also worry about the risk of professional embarrassment if she is misunderstood by laypeople (or other scientists) if she says that the uses are literal. This is a genuine worry, but it is no worse than other public misunderstandings of science.

As ISQ #7 points out, a poll on the question "Do neurons (literally) prefer?" is highly likely to yield an overwhelming consensus opinion of "No" among neuroscientists, not just laypeople. But this is *not* the same as asking them "What does 'prefer' refer to in neuroscience?" Answers to the poll question almost certainly depend on an implicit background conception of what "prefers" refers to when used for humans. Neuroscientists have internalized the same anthropocentric standard for the reference of "prefers" just like everyone else. In short, the results of asking the poll question will be informative about our common presupposed semantics of psychological terms, not about neuroscientists' commitment to a specific theory of their semantics.

In contrast, the reference question targets the issue directly. Yet if this question were asked instead, the most common answer would almost certainly be "response selectivity". In response, go to ISQ #2. In short, I would not expect neuroscientists' responses to the reference question to end the current debate either. They are also unlikely to provide a better account of literal use than linguists (see Chapter 6).

The point is not that neuroscientists don't hold relevant metaphysical or semantic views. For example, a behaviorist neuroscientist of the radical Skinnerian or Rylean logical varieties will say that "prefers" used for neurons refers only to a pattern of behavior, because that's all that psychological terms ever refer to. This is compatible with Literalism, albeit in a behaviorist or cognitive anti-realist version. To oppose Literalism, one must draw a *distinction* between human and nonhuman uses (see Chapter 7 for further discussion). A cognitive neuroscientist is in a more complicated, if more intuitively acceptable, position. She is independently committed to cognitive inferences from behavior. But her view of the extent of the cognitive is guided by the available evidence. If pressed, she might say that we do not have sufficient empirical justification to accept a version of (NA) or (NA-SC) for neurons and preferences. We don't have a mathematical model of this psychological construct—as I noted at the start of this section. So, if pressed, she may not be committed to the truth of sentences (1)–(3) above. But even this agnosticism does not rule out a Literal interpretation of "prefers" in (1)–(3). The term has sufficient sense for her research purposes to be part of fact-stating discourse. Even if its reference is under revision, it can still point to the same region of semantic space when used for humans and nonhumans. ("Gold" was not retired from use while it was undergoing reference revision.) If she wishes, she can assume the same methodologically behaviorist attitude towards neural preferences that many economists take towards human preferences: their research subjects are treated as black boxes. Anthropocentric tradition sanctions the additional cognitivist assumption that there are real preferences inside the human black boxes but not the neural ones. This tradition lies behind neuroscientists' probable answers to the poll question. But this tradition is what the Literalist questions. The prudent scientific course is to wait and see, and in the meantime to use the terms responsibly—that is, when the evidence indicates that neurons are exhibiting preferences, and not otherwise.[17]

The ISQ finally turns to what she may consider a *reductio ad absurdum*:

(#8)   But this opens the door to all kinds of abuses of language! If the stimulus-response profile of a neuron indicates *preferring*, why not say its readiness potential indicates its *eagerness* to fire? Why not say electrons *prefer* their orbitals?[18]

Why not, indeed? The objection seems to presuppose, falsely, that *Literalism* has opened the barn door (to elaborate ISQ #8's metaphor) to linguistic abuse. Such power!

---

[17] There is plenty of anecdotal evidence of openness to Literalism. Michael Graziano, discussing his book *Consciousness and the Social Brain* on Brain Science Podcast #108 (<http://www.brainsciencepodcast.com>), describes what he calls the emergent consciousness view, which he contrasts with his own theory: "Essentially, consciousness in that view is basically a magic property that emerges somehow, perhaps when information becomes very complicated, or the brain learns so much or gains so much memory that it crosses some threshold... Of course there's problems with that... We have a lot of really complicated information processing devices out there, like the Internet, and *I don't think the Internet has shown signs of being conscious yet*" (my italics). In other words, it's an empirical matter. Further anecdotal evidence is provided in Chapter 5, where neuroscientists who appear to endorse Bennett and Hacker's mereological fallacy in fact deplore psychological ascriptions to brains because there is insufficient evidence to support the ascriptions.

[18] My thanks to David Chalmers, Mark Sprevak, Mikaela Akrenius, and Kirk Ludwig for pressing this objection, albeit in somewhat different formulations.

The sad (to some) fact is that the barn door is *always* open and always has been—as Wilson (2006) also emphasizes. The interesting question is why our concepts so often remain in or close to the barn. How do we control and stabilize our conceptual behavior to the extent that we do? This question makes clear why the Literalist takes seriously the contexts and ways in which the extensions of psychological concepts are being made. They are contexts of articulation and confirmation of scientific theories of natural phenomena. Popular or pedagogical science and philosophical thought-experiments do not push these concepts out of the barn because they tell us where we have been conceptually, not where we are going. Merely using a term in a new way or in a new domain does not extend its explanatory scope or put pressure on its semantics; one needs the right kind of authority and context to make it stick.

That is why it is irrelevant that Dennett (or anyone) can appear to explain the behavior of a lectern by ascribing beliefs and desires to it. No one seriously thinks lecterns behave, nor does Dennett himself (Dennett 1981/1997: 66). As noted, there is no independently motivated explanandum here for which an intentional explanation might be a candidate explanans. The Intentional Stance is not a joke—I discuss it at length in Chapter 7—but the ascription of beliefs to lecterns is a bit of one. In contrast, biologists do seriously think that plants behave—a genuine conceptual leap that prior to being made hampered scientific progress (in the eyes of some). Some also argue that plant behavior is best explained in cognitive terms—and not in the amusing sense in which if you live in certain locales "you should plant apple varieties that are particularly *cautious* about *concluding* that it is spring—which is when they *want* to blossom, of course" (Dennett 1981/1997: 65). The peer-reviewed biological literature is evidence that the relevant concepts are under revision, albeit not nearly enough to include the domain of furniture.

It might be tempting for me to consider an additional condition for Literal ascription along these lines: is the psychological predicate being ascribed to an object that is generally considered to behave? We accept that animals and cells behave; we came to accept that plants behave; we don't accept that lecterns behave. However, I will resist this temptation. The ISQ can rightly point out that the term "behave" (and ditto for "evolve", "seek", "sense", and others) is already regularly and consistently used for physical phenomena generally, not just biological systems. Of course, I can point to different senses of "behaves". For example, Dretske

(1988: 11) defines behavior as internally produced movement by a system that has "enough structural complexity and internal articulation to make the internal-external difference reasonably clear and well motivated". For entities such as stones, lint, and drops of water, "behavior no longer means what it does with animals, plants, and more highly structured inanimate objects.... [Behavior] contrasts with what happens to an object, plant or animal. This is not the way the word 'behavior' is always used in (say) physics and chemistry" (Dretske 1988: 11).

But the ISQ might say: so what? Even if no one thinks an entity behaves in the sense appropriate to animals and other internally complex entities, what if a cognitive model were extended to, say, a body of water? This scenario would not be like Dennett's ungrounded extension of "belief" to a lectern. If, as I have argued, modeling practices can yield extensions of psychological predicates to any entity to which a model applies, then it seems I must accept these ascriptions as Literal.

I don't need to rely on distinguishing senses of "behavior" to respond to this form of ISQ #8. I'll flesh out the hypothetical case to show why.

Suppose an extension of Ratcliff's drift-diffusion model of decision-making to a body of water resulted in psychological ascriptions to the water. Specifically, scientists report to their colleagues that a body of water has been trained to distinguish specific stimuli and that in test situations it differs in its response times and accuracy, depending on how noisy the stimuli are, in measurable ways that fit Ratcliff's model. They report that the water made decisions and chose one stimulus over another. (Set aside the issue of replication.) The implausibility of the scenario does not detract from its metaphysical possibility. For the sake of argument, I'll even grant that it is empirically possible.

If Ratcliff's model were used in this way, it would indeed follow that we have relevant scientific evidence that the body of water makes decisions. Of course, the fact that the body of water isn't alive and lacks internal complexity would be legitimate scientific reasons to hesitate endorsing the ascription as *ultima facie* on this basis alone. I remind the ISQ that cognitive models are not developed in isolation. Biological background knowledge grounds our relatively ready acceptance of cognitive ascriptions to biological cases, and helps motivate the distinction in senses of "behavior" articulated by Dretske. Maturana and Varela's autopoiesis concept is another way of drawing this background distinction, and provides a theoretical limit to the extension of cognition.

Nevertheless, even if water does not behave in Dretske's primary sense or exhibit autopoiesis—starting with the basic fact that it does not maintain its own spatial boundary—it remains possible that these background distinctions will someday be erased or considered irrelevant, and the model-based extension of decision-making to water is accepted as *ultima facie*.[19]

But this challenge is exactly like imagining that the Internet is conscious (see fn. 17): both possible extensions depend on finding the relevant scientific evidence. Our affective responses to these imagined empirical results (i.e. ridicule or amusement) are evidentially irrelevant. So if the fundamental objection in ISQ #8 is that the Literalist cannot provide an a priori reason for ruling out lectern beliefs or water decisions, her answer is: there is none. Who knows? Maybe the Original Image will turn out to be correct after all. Science is no stranger to strangeness.

At this point, the ISQ tends to start repeating the same basic objections in slightly different terms, often by invoking various new or additional conditions—being able to reason self-reflectively about one's reasons, possessing self-awareness, and so on—that build into psychological concepts highly intellectualized features that they need not possess even to the folk. Using an ideal hyperrational, baroquely complicated human cognizer as the standard for real psychological capacities yields an emotionally satisfying chasm between us and other living things, but it also leaves most of us on the wrong side of the chasm most of the time.

4.4 Concluding Remarks

So far in this book, I have presented my positive case for Literalism. I have provided a variety of scientific cases that any theory of the semantics of psychological predicates must explain and that Literalism neatly explains. The default interpretation of non-psychological discourse in the relevant scientific contexts by the relevant experts is that it is intended literally to refer to natural phenomena. Literalism holds

---

[19] While Dretske's and Maturana and Varela's distinctions require internal complexity and therefore meeting an internal/external boundary condition, neither is restricted to biological entities (e.g. Beer 2015 emphasizes that "self-creating" is not restricted to living things). My conclusions would apply to extensions of psychological predicates to artificial agents.

that this default interpretation extends to psychological predicates in these contexts.[20] These predicates are systematically and selectively introduced via qualitative and quantitative analogy, and used to report and theorize about empirical results and future research. The uses are neither occasional nor accidental in motivation or employment. Literalism also provides a plausible and standard rationale for the spreading uses throughout biology. They are rational responses to new empirical findings. An adequate semantics for psychological terms should make sense of the trend, not just a few isolated predicates in one domain, one paper, or even one sentence. Literalism does so.

Literalism is also the cogent conclusion of an updated inference to the best explanation to other minds. It is defensible in the face of many intuition-driven objections. Finally, it coheres with an independently motivated semantic theory, although I will not show this until Chapter 6. An adequate semantics for psychological terms should either rely on a broader semantic theory, or else a new semantic theory employed for this purpose should not be restricted to dealing with these new uses alone. If it is, one would need to provide a strong reason to adopt one semantic theory for psychological predicates in nonhuman domains and another for everything else.

These explanatory virtues are worth keeping in mind as the relative merits of the alternatives to Literalism are assessed in the following chapters. For lay intuition, including philosophically enhanced lay intuition, may remain stubbornly on the ISQ's side. "Prefers", like most psychological terms, has strong human connotations and associations with concepts of motivation and purpose that are themselves steeped in anthropocentric lore (Keeton 1967: 452; Andrews 2009: 60). In the next three chapters, I present my negative case for Literalism. This consists in showing how, naïve intuitions aside, none of the alternatives rivals Literalism in its explanatory power. I begin with the Nonsense view.

---

[20] This does not mean that biologists are default realists, only that psychological predicates are univocal within fact-stating biological discourse, whether this discourse is given realist or anti-realist truth (or success) conditions.

5

The Nonsense View

5.1 General Remarks

In this chapter and the next two I critically consider three basic semantic alternatives to Literalism: the Nonsense view, the Metaphor view, and the Technical view. The latter groups a number of behavioristic or otherwise deflationary reinterpretations (deflationary relative to the traditional human standard). I will show that these alternatives are either implausible or less plausible than Literalism. This constitutes my negative argument for Literalism. To bowdlerize Conan Doyle: when the alternatives are shown to be implausible, what is left, however surprising, is probably true.

Although I have no demonstrative proof that there are no other possibilities, these alternatives exhaust those I have ever come across, in writing or conversation. If there are others, they are very likely to be variations on a theme presented in one or another of these chapters.[1] The uses are either (i) nonsensical (meaningless) or (ii) sensical (meaningful). If they are sensical, the word forms either (iia) retain their old meaning but are used to point to a different meaning in these contexts, or (iib) they have different meanings. (i) is the Nonsense view, (iia) is the Metaphor view, and (iib) is the Technical view (of which I discuss two

---

[1] For example, one might try a Fictionalist account in which the uses of psychological predicates for nonhumans take place within a fictional context, in which such claims are true within the fiction (the way "Sherlock Holmes is a British detective" is true within the fiction of the Conan Doyle stories). This view might be grouped with the Metaphor view (Chapter 6) into a category of Figurative (or Poetic) views. No one has ever suggested a Fictionalist account to me, but it is implausible anyway. The peer-reviewed scientific contexts I focus on are paradigms of non-fiction. This is not to say that being an *anti-realist* about scientific discourse in general (which may be expressed as a Fictionalist view) is implausible. But anti-realism is compatible with Literalism, which holds that the uses are univocal across human and nonhuman contexts however one accounts for their truth or success conditions. Relatedly, a Fictionalist view on the ontological status of models in science (that models are fictional entities) is also compatible with Literalism.

variants). The reason I've been able to identify a few basic alternatives is because many suggestions have in fact turned out to be variations on the same basic themes, usually (ii). I consider the sensical alternatives—the Metaphor and Technical views—in Chapters 6 and 7. In this chapter I consider the Nonsense view, defended by Bennett and Hacker (2003; Bennett et al. 2007). They argue that uses of psychological terms beyond certain bounds are nonsense, and that these uses of psychological terms are out of bounds.

5.2 The Nonsense View

Bennett and Hacker (2003; Bennett et al. 2007) take issue with uses by cognitive neuroscientists of psychological predicates to describe capacities of brains or parts of brains.[2] Neuroscientists "commonly try to explain human beings' perceiving, knowing, believing, remembering, deciding by reference to parts of the brain perceiving, knowing, believing, remembering, and deciding" (Bennett et al. 2007: 154). Prominently displayed excerpts include the following:

What you see is not what is really there; it is what your brain believes is there.... Your brain makes the best interpretation it can according to its previous experience and the limited and ambiguous information provided by your eyes.   (p. 16, from Crick 1995)

We seem driven to say that such neurons [that respond in a highly specific manner to, e.g., line orientation] have knowledge.... [T]he brain gains its knowledge by a process analogous to the inductive reasoning of the classical scientific method. Neurons present arguments to the brain based on the specific features that they detect, arguments on which the brain constructs its hypothesis of perception.   (p. 16, from Blakemore 1977)[3]

---

[2] I will draw citations from Bennett et al. (2007), which abridges critical passages in Bennett and Hacker (2003) by reproducing the relevant text from the latter in the former. Replies by Searle and Dennett are conjoined to this abridged version to form the text of Bennett et al. (2007). Hacker's reading of Wittgenstein as a logical grammarian, concerned with rules in ordinary language that fix clear bounds of sense, is a standard one, but alternative readings abound. In any case, Hacker's reading grounds the Nonsense view.

[3] More examples are provided on pp. 154–6, as well as sprinkled throughout Bennett et al. (2007) and Bennett and Hacker (2003). These examples are representative of many of the uses they directly target and the *Scientific American* or TED-Talk-like level of exposition in which the uses figure (albeit in the pre-TED-Talk era).

While the scientists they excoriate are among the most prominent—Francis Crick, Antonio Damasio, and Gerald Edelman, to name a few—the sources from which they draw their primary examples are popular science books, including transcribed radio or television broadcasts, that are aimed at lay (non-neuroscientist) audiences, including philosophical audiences, as in Zeki (1999). Often they are somewhat dated in neuroscience terms (e.g. Blakemore 1977; Young 1978), although at least some recent peer-reviewed work in cognitive neuroscience (discussed below) involves similar usage (or misusage).[4] Nevertheless, their starting point, unlike the Literalist's, is contexts of science communication and public outreach.

Bennett and Hacker's overall view of concepts informs their interpretation of these uses of these concepts and their responses to various sensical alternatives that they reject. Concepts have rules that determine their proper use, and "[n]onsense is generated when an expression is used contrary to the rules for its use" (Bennett et al. 2007: 12). When Freud and others introduced the concepts of unconscious beliefs, desires, and motives by analogical extension, the new uses involved "a different, importantly related, meaning"—minor rule revision within the same language game. In the neuroscience cases, however, "these psychological expressions have not been given a new meaning"—there are no new rules being provided to establish new concepts. Still operative are the old rules establishing the bounds of sense of "old non-technical concepts— concepts of mind and body, of thought and imagination, of sensation and perception, of knowledge and memory, of voluntary movement and of consciousness and self-consciousness" (Bennett et al. 2007: 11). The transgression of these rules "motivates our claim that neuroscientists are involved in various forms of conceptual incoherence" (Bennett et al. 2007: 29). As they put it:

[Our motivation] is rather the acknowledgement of the requirements of the logic of psychological expressions. Psychological predicates are predicable only of a whole animal, not of its parts. No conventions have been laid down to

---

[4] They also take aim at concepts of representation and rules used in computational or information-processing models of cognition (Bennett et al. 2007: 155–6). These concepts were the focus of a debate between Pylyshyn (1993) and Boyd (1993) as to whether they are literal (Pylyshyn) or metaphorical (Boyd) when used in these models. Pylyshyn and Boyd agree the terms are sensical.

determine what is to be meant by the ascription of such predicates to a part of an animal, in particular to its brain. So the application of such predicates to the brain or the hemispheres of the brain transgresses the bounds of sense. The resultant expressions are not false, for to say that something is false, we must have some idea of what it would be for it to be true—in this case, we should have to know what it would be for the brain to think, reason, see and hear, etc. and to have found out that as a matter of fact the brain does not do so. But we have no such idea, and these assertions are not false. Rather, the sentences in question lack sense. This does not mean that they are silly or stupid. It means that no sense has been assigned to such forms of words, and that accordingly they say nothing at all, even though it looks as if they do.

(Bennett et al. 2007: 29–30)

What they call the "mereological fallacy" follows from the fact that the new uses are not licensed by the old rules. The fallacy is a logical or conceptual mistake. It involves using a predicate that, given its rules for proper use, denotes a property of and only of a whole in order to (try to) ascribe that property to a part. Psychological terms are properly applied only to whole humans, and perhaps some other animals, but certainly not to their parts. As they put it (Bennett et al. 2007: 149), "We no more understand what it would be for a brain or its parts to think, reason, fear, or decide something than we understand what it would be for a tree to do so." Note, however, that if the rules forbid psychological ascriptions to trees, this cannot be a case of the mereological fallacy.

This fallacy is "remediable by a correct account of the logico-grammatical character of the concepts in question" (Bennett et al. 2007: 11). Wittgenstein is cited as the authority for this view:

[O]nly of a living human being and what resembles (behaves like) a living human being, can one say that it has sensations; it sees, is blind; hears, is deaf; is conscious or unconscious. (Wittgenstein 1958: §281; Bennett and Hacker 2003: 19)

("This [remark] epitomizes the conclusions we shall reach in our investigation": Bennett et al. 2007: 19.) On this interpretation of Wittgenstein's views of ordinary language, human behavior is criterial for the proper use of a psychological term; it is not just a reliable basis for inductive inference to the capacity it denotes. It is a conceptual mistake to use psychological terms for brain parts (unless the uses are figurative, as in metaphor, metonym, or synecdoche) because brain parts *cannot* behave in a way that resembles the behavior of a living human being: "it makes no sense to ascribe such psychological attributes to anything less than the animal as a whole" (Bennett et al. 2007: 7).[5] One can sum up the view as follows:

(1) Behaving like a human is criterial for applying psychological predicates.
(2) If X can't behave like a human, then it is a conceptual mistake to apply psychological predicates to X.
(3) Brains or brain parts can't behave like humans.
(4) So it is a conceptual mistake to apply psychological predicates to the brain.
(5) So when they are so applied, the result is nonsense.

They add that the problem is not just that cognitive neuroscientists are misusing words and spreading confusion. It has important ramifications:

Human beings possess a wide range of psychological powers, which are exercised in the circumstances of life, when we perceive, think and reason, feel emotions, want things, form plans and make decisions. The possession and exercise of such powers define us as the kinds of animals we are. (Bennett et al. 2007: 6)

The conditions in which we satisfy the criteria for correct, literal, application of psychological terms are also the conditions in which we display the powers that define us as human. The possession of these capacities is what makes a human being a person, where being a person is "roughly speaking, to possess such abilities as would qualify one for the status of a moral agent" (Bennett et al. 2007: 134). So the misuses in cognitive neuroscience blur important moral as well as ontological distinction. The latter distinction appeared in Chapter 1 in terms of the restriction of person-appropriate predicates to humans that yielded the Manifest Image. The former exemplifies the traditional link between the possession

---

[5] This is not wholly clear. For example, they note (Bennett et al. 2007: 135): "[T]he *concept* of consciousness is bound up with the behavioral grounds for ascribing consciousness to the animal. An animal does not have to exhibit such behavior in order for it *to be* conscious. But only an animal to which such behavior *can intelligibly be ascribed* can also be said, *either truly or falsely*, to be conscious." But if one can intelligibly ascribe consciousness to an animal (or human) that does *not* exhibit the behavior—as in cases of minimal consciousness, for example—then the grounds for ascription need not be behavioral, and the claim that the concept is "bound up" with the behavioral grounds in the constitutive, logical sense needed for the Nonsense view (Bennett et al. 2007: 127, 149) needs justification. Searle avoids this issue because he thinks consciousness arises from the brain; for him, the concept of consciousness will not be "bound up" or constitutively grounded in behavior.

of psychological capacities and the moral status of entities that possess these capacities. I discuss this link in Chapter 9.

5.3 The Literalist Responds

In a critical sense Bennett and Hacker and I are writing at cross-purposes. Their ire is aimed at cognitive neuroscience, and their primary examples are largely taken from works intended for popular audiences. My concern is the uses of psychological language throughout biology, and my primary examples are from peer-reviewed publications, including Nobel prize-winning papers, written for other scientists. Peer-reviewed publications, not popular science, are the primary means by which empirically-supported assertions about natural phenomena are widely disseminated among scientists.

These differences make a difference. Bennett and Hacker are correct to upbraid many cognitive neuroscientists for occasionally playing fast and loose with psychological predicates in their popular writings. But they do not even begin to engage with the motivating reasons for Literalism: the broad range of new domains of systematic use, and the extension of psychological predicates by means of models. So while their position reflects a standard semantic theory (however unpopular that account might currently be) they do not begin to address the questions of systematic plausibility and new types of evidence. That may never have been their intention, of course. But a straightforward extension of their view entails that as biological knowledge has increased, scientists have become increasingly nonsensical in describing their discoveries. This positive correlation of more knowledge and more nonsense is implausible. Even if we restrict the discussion to cognitive neuroscience, their view entails that Nobel prize-winning neuroscientists are writing nonsense in the papers that helped garner them the prize.

Because they do not engage with the relevant scientific literature, they miss the reasons that show their interpretation is mistaken even by their own lights. The clearest way to show this is by examining their rejection of the possibility that neuroscientists are extending the psychological vocabulary analogically:

It is indeed true that analogies are a source of scientific insight. . . . The moot question is whether the application of the psychological vocabulary to the brain is to be understood as analogical. The prospects do not look good. The application of psychological expressions to the brain is not part of a complex theory replete with functional, mathematical relationships expressible by means of quantifiable laws as are to be found in the theory of electricity.   (Bennett et al. 2007: 28–9)

Whether "laws" are necessary for scientific theory is not clear. This logical positivist view of theories was widely rejected long ago, in part because respectable biological theories often didn't contain laws. But—no doubt inadvertently—in this comment Bennett and Hacker underline why the model-based extensions discussed in Chapter 3 put so much pressure on the semantics of psychological predicates via a non-anthropocentric understanding of the properties and capacities to which they refer. This is also revealed in their explanation of why the intelligible ascription of psychological predicates to the brain is a conceptual issue, not an empirical one:

We understand what it is for people to reason inductively, to estimate probabilities, to present arguments . . . But do we know what it is for a brain to see or hear, for a brain to have experiences, to know or believe something? Do we have any conception of what it would be for a brain to make a decision? . . . These are all attributes of human beings. Is it a new discovery that brains also engage in such human activities? Or is it a linguistic convention, introduced by neuroscientists, psychologists, and cognitive scientists, extending the ordinary use of these psychological expressions for good theoretical reasons? Or, more ominously, is it a conceptual confusion? . . . One cannot investigate experimentally whether brains do or do not think, believe, guess, reason, form hypotheses, etc. until one knows what it would be for a brain to do so, i.e. until we are clear about the meanings of these phrases and know what (if anything) counts as a brain's doing so and what sort of evidence supports the ascription of such attributes to the brain.   (Bennett et al. 2007: 18–19)

They go on to argue that it would be "astonishing" (Bennett et al. 2007: 20) to discover that it is not "only of a human being and what behaves like a human being" that psychological ascriptions are true; we would want to know what the evidence for the ascriptions was. They cite research by Susan Savage-Rambaugh in which trained bonobo chimpanzees "can ask and answer questions, can reason in rudimentary fashion, give and obey orders, and so on. The evidence lies in their behavior (including how they employ symbols) in their interactions with us." In contrast, "the neuroscientists who adopt these forms of description have not done so as a result of *observations* which show that brains think and reason" (Bennett et al. 2007: 20; their italics).

It is questionable whether it is appropriate to judge the cognitive capacities of nonhuman animals by how well they can imitate what humans do in human environments. Nor is it clear whether Bennett and Hacker intend to interpret "resembles (behaves like) a living human being" invariably in this fine-grained a manner. But the moot question is whether establishing qualitative similarity to human behavior is still our best method for determining the proper extensions of psychological predicates. We may not have functional, quantifiable laws, but we do have functional, quantifiable mathematical models to do the job of establishing relevant similarity to human behavior. Because of these models, we have clear answers to the questions posed in the passage just cited. We *do* have a conception of what it would be for a brain to make a decision. We *do* have reason to think there is linguistic innovation by neuroscientists, psychologists, and cognitive scientists for good theoretical reasons. We *do* know what would count as a brain's doing such-and-such and what sort of evidence supports the ascription of such attributes to the brain. In short, even if we take on board their Wittgenstein-inspired rule-based semantics, the rules for psychological predicates *have* changed. They have changed for humans and nonhumans alike. They now depend in part on the kind of quantifiable evidence that Bennett and Hacker *agree* would show that the brain makes decisions. The brain *would be* behaving like a human—just not in the qualitative, intuitive sense that Bennett and Hacker take for granted.

Of course, we also don't have to adopt their semantic theory. Wittgenstein is also widely interpreted as holding the view that there are no determinate rules governing the proper uses of many terms:

> For I can give the concept "number" rigid limits in this way, that is, use the word "number" for a rigidly limited concept, but I can also use it so that the extension of the concept is not closed by a frontier. And this is how we do use the word "game". (Wittgenstein 1958: §68)

The late Wittgenstein's linguistic open-endedness undermines both Bennett and Hacker's mereological fallacy and the Wittgensteinian justification of their rule-based semantics. For example, it is associated with concept introduction in science (albeit metaphorically to some: Boyd 1993: 482). Wilson (2006: 23) articulates examples from the history of science to argue that "there is no reason to expect our linguistic training . . . secretly anticipates the later adaptations in any reasonable sense", citing Wittgenstein's *Philosophical Investigations* in support. If Wittgenstein can be cited in arguments from authority by both sides, he is probably best left out of the debate altogether.[6]

For the fun of it, it is worth taking a few sentences from the peer-reviewed scientific literature (also cited in Ch. 4) and turning them into what is widely agreed to be nonsense (with apologies to Lewis Carroll and credit to Noam Chomsky):

(1) A resonator neuron *prefers* inputs having certain frequencies that resonate with the frequency of subthreshold oscillations of the neuron. (Izhikevich 2007: 3)

(2) In *preferring* a slit specific in width and orientation this cell [with a complex receptive field] resembled certain cells with simple fields. (Hubel and Wiesel 1962: 115)

(3) A resonator neuron *gyres* inputs having certain frequencies that resonate with the frequency of subthreshold oscillations of the neuron.

(4) In *gyring* a slit specific in width and orientation this cell [with a complex receptive field] resembled certain cells with simple fields.

(5) Colorless green ideas sleep furiously.

Sentences (1)–(4) must be nonsense—they say nothing at all, even if it looks as if (1) and (2) do—because "prefers" is the only term that could make (1) and (2) nonsense and "gyre" in (3) and (4) is independently agreed to be nonsense. (5) is Chomsky's famous example of a semantically nonsensical but syntactically correct sentence. Moreover, if the rules of use for "prefers" are so unequivocal, the first two sentences should strike all or most of us as equally nonsensical as the last three. Otherwise, we are entitled to an explanation of why Bennett and Hacker know the semantic rules of "prefers" better than the rest of us.[7] If they have in mind

---

[6] The linguistic openness to which Wittgenstein gestures via the concept of a game can be accommodated in linguistic theory. For example, in cognitive linguistics, Croft and Cruse (2004: 75 passim) propose that words have meaning potentials that occupy regions of conceptual space, rather than ready-made senses or rules for use. Clear meaning boundaries can be established in particular construals in context (2004: 95).

[7] In peer-reviewed contexts, neuroscientists assume that their readership is aware of how words are used in the relevant literature. This doesn't mean they have laid down rules for the terms that establish their use in all their contexts, however these meanings are related to non-scientific meanings. So Bennett and Hacker may be correct that in science communication contexts the neuroscientists are using words with their ordinary meanings to explain the science using terms the folk understand. But it should not be assumed that the account

some other notion of "nonsense", it would be helpful to know what this other sense might be and why it is relevant. So when they claim (Bennett et al. 2007: 128), "If a form of words makes no sense, then it won't express a truth", it is plausible to counter that (1) and (2) are expressions of truths (or empirically confirmed statements), so the Nonsense view fails.

As a final point, the mereological fallacy looks ad hoc even on their own terms, and approving appropriations of it in the scientific literature typically mistake its tenor. Bennett and Hacker acknowledge that the fallacy does not hold for all psychological predicates. Verbs of sensation (e.g. to hurt, as in "My hand hurts") are among the "striking exceptions" (Bennett et al. 2007: 133) in which ascriptions to parts is semantically legal. Why are they exceptions? Many non-psychological verbs are also "exceptions": opening, turning, living and dying, transferring, and other activities or processes are frequently ascribed at both whole and part levels without conceptual rule-breaking. In many ways, psychological verbs appear to behave just as verbs do in general (Figdor 2017).

I and others agree with Bennett and Hacker's critical attitude towards cognitive neuroscience regarding parts of that literature. Cognitive neuro-scientists are often not clear about what they mean, defining terms in orthodox behaviorist manner and then drawing inferences that presuppose a cognitive interpretation (Figdor 2013). This is a big problem for public communication of neuroscience (Bennett et al. 2007: 47; Illes et al. 2010; Figdor 2013). The more egregious of these uses ascribe psychological concepts to brain areas on the basis of correlations between brain activity (as measured by fMRI or other scanning technologies) and human behavior. This practice has been dubbed neo-phrenology (Uttal 2001; Shulman 2013). Such correlations are insufficient for inferring that a particular brain area is even necessary, let alone sufficient, for a cognitive capacity of a particular type. Behind these misuses is the difficulty of finding the right vocabulary for labeling the functions of brain areas or networks (e.g. Price and Friston 2005; Cacioppo et al. 2008: 66; Anderson 2010; Lenartowicz et al. 2010; Poldrack 2010: 754; Figdor 2011).[8]

But these concerns are empirical. Despite some approving references to the mereological fallacy in the neuroscience and psychological

---

we give for uses in science communication contexts is the same as the one we give for uses in serious scientific contexts. The debate in this book is only about the latter.

[8] This claim is not intended to exclude wholesale revision or eliminativism a priori.

literature, few if any notice its logical (or conceptual) nature. They embrace Bennett and Hacker's criticisms but not their semantics. For example, Miller (2010: 718 and passim) approves of their "lamenting the sloppy use of language in the naïve biological reductionism of equating biology and psychiatry". But Miller takes issue with identifying psychiatric diseases with biological conditions without having underlying causal-mechanistic accounts that would justify asserting biological-psychological identities. The sloppy talk involves empirically unjustified identity claims, not violated conceptual rules.

Similarly, Shulman (2013: 54–5) agrees with Bennett and Hacker's view that cognitive neuroscience is mistaken in ascribing psychological attributes to the brain. But the "conceptual transfer" that Shulman opposes is the idea that mind–body dualism is resolved when "the psychological properties formerly assigned to the mind are assumed to have been transferred to the brain, where—since the brain is material—they can presumably be studied scientifically". He adds that Bennett and Hacker think this transfer is wrong because it involves the mereological fallacy, "wherein a property of the whole is assigned to a part". But this is the fallacy of division, a metaphysical error (when it is an error). For Shulman, the transfer is wrong because as a behaviorist neuroscientist he is interested in the relationships between behavior and brain activity. From his perspective, it wouldn't matter if Bennett and Hacker's interpretation of the meanings of psychological concepts was wrong. In his review of Shulman (2013), Seth (2014) agrees with Shulman that brains are necessary for making decisions but do not themselves make them. But on his view this is because "behaviour and cognition emerge from rich interactions among brains, bodies, and environments and do not arise from neural activity alone" (Seth 2014: 5). For both Shulman (a behaviorist neuroscientist) and Seth (a cognitive neuroscientist), it is not *conceptually* ruled out that the brain cannot be properly ascribed decision-making. Such claims are mistaken for empirical reasons or from an independent commitment to a version of behaviorism.

Finally, in his look at competing conceptual frameworks for immune systems, Tauber (2013: 257) notes that, in a Gibson-inspired dynamical process view, cognition is an appropriate way to describe behavior of organisms, and that organisms are "not to be described in terms of the immune or nervous systems". Citing Bennett and Hacker (2003), he writes that "confusing these different kinds of descriptions is to commit the 'mereological fallacy'". But this is not Bennett and Hacker's mereological fallacy either. To Tauber, committing this mistake does not yield nonsense. On Tauber's view, cells and molecules comprising the immune system "are not themselves cognitive except as used metaphorically in their physical descriptions".[9]

In sum, the Nonsense view fails as a plausible alternative to Literalism. Bennett and Hacker's targeting of popular cognitive neuroscience misses the forest for some epistemically inconsequential bushes nearby. Even on their own terms, we have independent reason to think the uses in neuroscience are sensical based on quantitative analogy, a crucial source of evidence that they do not consider. The mereological fallacy appears ad hoc, and when it is apparently endorsed by neuroscientists its logico-conceptual nature is ignored or misunderstood. Finally, the debate is not well-served by turning it into a contest between competing arguments from Wittgensteinian authority.

5.4 Dennett and Searle Respond

Among the semantic "escape routes" (Bennett et al. 2007: 25) that Bennett and Hacker consider are the claims that the terms are new, technical homonyms of ordinary psychological predicates; that they are analogical extensions; and that they are figurative or metaphorical or even poetic. To block the first escape route, they reply that Crick, Zeki, and the rest are not using the terms in novel ways, given the inferences that they draw. The Literalist agrees. Their reply to the analogical-extension view was discussed above. In response to the third, they repeat the claim about the sameness of inferences, adding that it is dangerous not to make clear that the uses are metaphorical on pain of confusing old and new uses and generating incoherence. The Literalist agrees regarding the potential for confusion, but thinks it is generated by failing to clarify that human behavior is not criterial for their meaning (to put the point in their terms).

In their face-off with Bennett and Hacker, Dennett and Searle both choose a sensical, but non-Literalist, alternative to the Nonsense view.

---

[9] Tauber attributes this last comment to a personal communication. As I understand him, Tauber is not so much defending either position but rather presenting and comparing the two conceptual frameworks for characterizing immune systems.

These alternatives will be explored in the next two chapters, but discussing Dennett's and Searle's direct responses to Bennett and Hacker enables me to clarify the Literalist position further.

Dennett agrees with them about "unacknowledged Cartesian leftovers strewn everywhere in cognitive neuroscience and causing substantial mischief" (Bennett et al. 2007: 74). But he finds that their "remarkably insulting attack on me" (Bennett et al. 2007: 75)—or rather, Hacker's attack (Bennett et al. 2007: 77)—ignores the fact that he made the same point about subpersonal uses of cognitive terms way back in 1969, where he introduced the personal/subpersonal distinction (Dennett 1969). There Dennett made clear, for example, that "talk of pains is non-mechanical, and the events and processes of the brain are mechanical" (Dennett 1969: 91).[10] But for Dennett, rather than conceptual confusion, the uses of psychological terms for brain parts (among many other nonhuman entities) reflect both new discoveries and linguistic innovation (Bennett et al. 2007: 86). The personal/subpersonal distinction also appears in homuncular functionalism as the distinction between the ascription of full-fledged cognitive properties to wholes and the ascription of "attenuated" or "hemi-semi-demi-proto-quasi-pseudo" cognitive properties to their parts (Bennett et al. 2007: 88). The mereological fallacy just is the homuncular fallacy—the epistemically damaging fallacy of ascribing full-fledged cognitive properties to parts (Bennett et al. 2007: 78).

Thus, Hacker's lambasting Dennett for failing to appreciate the mereological fallacy "is a case of teaching your grandmother to suck eggs" (Bennett et al. 2007: 77). On his view, the mereological/homuncular fallacy is not a fallacy (Bennett et al. 2007: 87)—or, better, it *is* a fallacy, but neuroscientists and many other scientists are not committing it. This is because they *aren't* ascribing full-fledged properties to the parts. Cognitive scientists have their own semantic rules, and they differ from the rules Bennett and Hacker think are still in play when psychological terms are used at subpersonal levels. For Bennett and Hacker, the meanings do not differ at the brain level. The same semantic rules are operative for ascriptions to parts, they're being violated, and the result is

---

[10] Bennett and Hacker respond that their distinction is not the personal/subpersonal distinction. Their reason is that the mereological fallacy does not involve a non-mechanical/mechanical distinction (Bennett et al. 2007: 132).

nonsense. For Dennett, the meanings *do* differ at the subpersonal levels. The semantic rules are different for ascriptions to parts. So while he agrees with Bennett and Hacker that uses of (e.g.) "believes" at the subpersonal level are not literal *qua* ascriptions of full-fledged belief, for Dennett they are literal, hence sensical, *qua* ascriptions of partly-fledged belief. This proposed interpretation of the terms as having attenuated referents at subpersonal levels—a result of "the poetic license granted by the intentional stance" (Bennett et al. 2007: 89)—will be critically examined in Chapter 7.

Dennett also finds that the passage Bennett and Hacker quote from Wittgenstein as the source of their Nonsense view is ambiguous:

I am happy to cite this passage from Wittgenstein myself; indeed I take myself to be extending Wittgenstein's position: I see that robots and chess-playing computers and, yes, brains and their parts *do* "resemble a living human being (by behaving like a human being)"—and this resemblance is sufficient to warrant an adjusted use of psychological vocabulary to characterize that behavior.

(Bennett et al. 2007: 78)

I made a similar point above, although Dennett and I part company over whether the resemblance involves *doing what humans do from the lay perspective* or *doing what humans do from the model-based science perspective*. This difference informs our difference in how we interpret psychological predications at the personal and subpersonal levels. On my view, they ascribe the real thing at subpersonal levels, with nothing attenuated or pseudo about them. Naturally, this raises red flags about Literalism and the status of the homuncular fallacy. I address this issue in Chapter 8.

Searle's response to Bennett and Hacker is to take aim at the mereo-logical fallacy as a new example of a category mistake (Bennett et al. 2007: 107). The logical error follows from "the Wittgensteinian vision: because the conscious behavior cannot be exhibited by the part, i.e. the brain, and because the conscious behavior is essential to the attribution of the consciousness, we cannot attribute pains to the brain" (Bennett et al. 2007: 106-7). Searle denies that behaving like a human being is criterial for *all* proper applications of psychological predicates. He argues that we can ascribe cognitive capacities to the brain as agent or subject ("the brain thinks"), or as locus of cognitive activity ("thinking occurs in the brain") (Bennett et al. 2007: 107). At most, the criterial link between human behavior and psychological predicates would apply to ascriptions to the "brain as subject", but not to ascriptions to the "brain as locus". The latter ascriptions are what we observe in cognitive neuroscience. These uses are "harmless metaphors" (Bennett et al. 2007: 112–14), although his suggestion that they are metaphorical is not elaborated. Instead, Searle focuses on why the ascriptions sound odd even though there is nothing wrong with them.

The reason it sounds odd, but is perfectly sensical (if metaphorical), to attribute decision-making to the brain as locus is because we typically "carve off the social situation and identify a purely psychological component" (Bennett et al. 2007: 120). To use his sentences (renumbered here):

(6)   I can visually discriminate blue from purple.
(7)   I have decided to vote for the Democrats.
(8)   I own property in the city of Berkeley.

Searle argues that "My brain" can substitute for "I" in (6) and (7); Bennett and Hacker reject both substitutions. With the substitution, (6) becomes a straightforward case of a psychological ascription to the brain as locus. In (7), we "carve off" the social situation of the ascription when we make the substitution; this explains why saying that my brain decided to vote for the Democrats *sounds* odd even if it is still a brain-as-locus ascription. In (8), we cannot "carve off" anything because there is nothing we can attribute to the brain alone (Bennett et al. 2007: 121). The property owner is the embodied brain, but only under social and legal aspects can it be a property owner.

In any case, neuroscientists need not worry about whether the brain should be ascribed rational agency. They are perfectly fine investigating the brain and ascribing psychological properties to the brain as locus. The oddness of psychological ascriptions to the brain has a non-semantic explanation in terms of the "carving off" of social contexts of ascription.[11] For Bennett and Hacker, the oddness indicates semantic rule-breaking.

---

[11]   This may be understood as a semantic explanation if the social aspects are part of the meaning of the psychological term "decides" when it is used in (7), and we somehow carve them off to reveal a non-social "core" meaning. However, I take it that Searle's main point is to argue that the felt oddness of saying the brain decides does not entail Bennett and Hacker's explanation of it in terms of the violation of logico-conceptual rules.

For the Literalist, it stems from a felt conflict with our anthropocentric semantic tradition for psychological terms.

5.5 Concluding Remarks

Bennett and Hacker's condemnation of the confusing way psychological predicates are used in popular cognitive neuroscience is widely embraced; the reasoning behind their condemnation, not so much. Few accept that the uses should be interpreted as nonsensical. For the Literalist, the devastating flaw in the Nonsense view is that it completely misses the scientific challenge to our traditional anthropocentric grounds for determining the proper extensions of psychological predicates. Even if we take on board Bennett and Hacker's account of meaning in terms of semantic rules, the rules are changing because of the use of quantitative models of cognition. Our evidence for what counts as "behaving like a human being" is changing. As a result, the criterial status of human behavior is also fading. Literalism takes these changes seriously and offers a plausible way to understand psychological ascriptions in their light.

For Dennett and Searle, the semantic rules also allow sensical, albeit attenuated or metaphorical, psychological ascriptions to parts of humans. Both of these semantic alternatives to Literalism and the Nonsense view are generalizable to the uses of psychological predicates across biology. I consider them in turn in Chapters 6 and 7, starting with the Metaphor view.

6

The Metaphor View

6.1 General Remarks

When DasGupta et al. (2014) write that fruit flies *decide*, and when Hubel and Wiesel (1962) write that neurons *prefer*, a popular initial response to these unexpected uses is that they are intended metaphorically. The Metaphor view claims that the uses make sense (*pace* Bennett and Hacker) but aren't literal (*pace* Literalism). To the best of my knowledge, a defense of the Metaphor view has not appeared in print. Besides Sellars, Searle (in Bennett et al. 2007) and Dennett (2009: 8; 2013) also suggest the uses are metaphorical without elaboration. In this chapter, I will sort through several metaphorical views and show that none is a plausible alternative to Literalism.

My main target is the Metaphor view as a semantic proposal (denoted with a capital "M"). This view holds that psychological terms in the relevant scientific contexts still have their conventional meanings, but are being used metaphorically in these contexts. As Sellars (1963/1991: 11–12) expresses the view:

[T]he use of the term 'habit' in speaking of an earthworm as acquiring the habit of turning to the right in a T-maze, is a metaphorical extension of the term. There is nothing dangerous in the metaphor until the mistake is made of assuming that the habits of persons are the same sort of thing as the (metaphorical) 'habits' of earthworms and white rats.

Like the Nonsense view, it is a conservative strategy. If either view is correct, the predicates refer to what they always have, by our own anthropocentric lights. Both views imply that the reference of psychological terms is not being revised in the light of scientific discovery. There is no metaphysical problem, only an obvious semantic one.

Although I will not press this point further, it is worth noting from the start that the Metaphor view is systematically implausible. Just as the

Nonsense view implausibly implies that more knowledge positively correlates with more peer-reviewed nonsense, the Metaphor view implies that more knowledge positively correlates with more peer-reviewed poetry.[1] Biologists are unlikely to be this irrational. For the Literalist, of course, they are acting perfectly rationally. In any case, in what follows I will discuss individual sentences in isolation, setting aside the significance of the broad context of systematic, consistent, and empirically-driven use across biological disciplines.

The Metaphor view makes two core claims: that psychological predicates refer to humans-only capacities in their proper, literal, uses, and that the referential intent of scientists when they ascribe these capacities to nonhumans is metaphorical. The phrase "humans-only" is shorthand for the idea that the proper extensions of these predicates are very restricted. A plausible Metaphor view can and should hold that literal uses may include the usual nonhuman suspects (chimpanzees, dolphins, and so on). What they do not include are entities from the broad range of nonhuman kinds now (apparently) being ascribed the relevant properties or capacities. I will target each of these core claims. First, I find no independent evidence for the claim that the terms are being used with metaphorical intent. Showing that they are is far more difficult than one might think. Second, leading theories of semantic metaphor do not provide independent support for the idea that the meanings of psychological predicates rule out literal uses in the unexpected domains.

In addition to denoting a semantic position, the term "metaphor" is also often used to indicate scientific practices of theory building via analogical reasoning (Hesse 1966; Gentner 1983; Gentner and Jeziorski 1993: 453; Medin et al. 1993; Gentner and Markman 1997; Gentner et al. 2001: 200). I call this epistemic metaphor. I discuss it only briefly below, as it is obviously consistent with Literalism. A term extended by analogy—e.g. "orbit", to electrons—can of course be intended literally.

As a final preliminary matter, I take the Metaphor view to be compatible with whatever theory turns out to be best regarding the nature of metaphorical processing. This includes in particular the question of whether metaphor is fundamentally linguistic or psychological (conceptual). My

---

[1] Unfortunately, as more scientists engage with the lay public (and as more information and misinformation is available to the public), there will be more opportunities for public misunderstanding of science in potentially problematic ways (e.g. Pauwels 2013).

concern is with the relation between mind or language and the world, not between mind and language. For example, the structure-mapping theory (e.g. Gentner and Markman 1997; Gentner et al. 2001) holds that metaphors are formed by establishing an alignment or mapping between structured representations (concepts) of a source object or situation, including its properties and relations, and a target object or situation and its properties and relations.[2] The theory tells us how the similarity mapping is made. That the representations refer to distinct things is taken for granted. Here, the issue is the relation between the referents of the representations. It is whether what these representations are about when they are used for nonhumans is the same as what they are about when used for humans. As a result, it will not matter if the discussion proceeds in terms of conceptual or linguistic representations (concepts or words), and whether metaphor is fundamentally linguistic or conceptual. It is agreed that psychological concepts or words refer literally in the case of humans, and the issue is whether the uses for nonhumans are also literal or else metaphorical.

A further complication is the fact that it is standard to hold that the meaning of a word is the concept it expresses (or is the concept it is used by a speaker to express on an occasion of use).[3] In many contexts the terms "concept" and "meaning" are used interchangeably (e.g. Carey 1988: 167 fn. 1). For example, Bennett and Hacker summarize their work as an effort to show that "clarity concerning conceptual structures" is critical in cognitive neuroscience (Bennett et al. 2007: 48). Dennett adds that the philosopher, who is "an expert on nuances of meaning", is the right sort of person "to conduct this important exercise in conceptual hygiene" (Bennett et al. 2007: 74).

---

[2] Common features can include similarity of appearance, action or capacity, function, and so on. Which features we choose to compare, how we alight on a particular target for comparison, and how we judge similarity between source and target are matters of ongoing cognitive psychological research (see, e.g., Glucksberg and Keysar 1990; Bowdle and Gentner 2005; Camp 2006; Nersessian 2008). Other prominent debates not of concern here include the general kinds of information in concepts (prototype, exemplar, and theory theories), their representational format (modal vs. amodal), how concepts come to be possessed (nativism vs. empiricism), and whether concepts, understood as mental representations, are needed in cognitive science.

[3] While for some (e.g. Davidson 1984) meaning and truth conditions (which depend on reference) are identical, I assume the dominant view that truth conditions (and reference) do not exhaust meaning, which also depends on other conventions and on what may be contextually included on an occasion of use. This is discussed further below.

In consequence, for the most part I too will use "concept" and "meaning" interchangeably. I'll write that words and concepts refer, and I won't distinguish between a word or concept's referring and its being used by a speaker to refer on an occasion of use. Relevant nuances are considered in more detail when I discuss classical Gricean and relevance-theoretic accounts of linguistic meaning and metaphor. I adopt the conventions of using capital letters to denote concepts (e.g. PREFERS) and mention quotes (never scare quotes) to mention words (e.g. "prefers").[4]

6.2 The Metaphor View

As might be inferred from the above, metaphor is complicated. Thankfully, the Metaphor view itself is simple. When terms conventionally used to express psychological concepts (and ascribe psychological capacities) are used by scientists to describe nonhumans, the Metaphor view claims that we should interpret them as using the terms metaphorically. A metaphorical use of a term involves using it to express a concept other than the concept it customarily expresses in literal use. For many theorists, this is done by drawing a similarity mapping between features of the terms' (or concepts') referents.[5] For example, suppose a couple lovingly watches their toddler play in a mud puddle. The husband says to his wife:

(1) Our piglet is getting dirty. (Bezuidenhout 2001: 161)

The term "piglet" is being used metaphorically to refer to the child by way of an activity common to toddlers and piglets that their toddler is happily engaging in. The term conventionally expresses the concept

---

[4] The idea that concepts are meanings is distinct from the idea that concepts are (realized as) information-bearing states (e.g. Field 1973). I use the term "concept" for meanings of terms and "mental representation" for what many intentional realists claim about their realization. Thus, saying that psychological predicates pick out psychological states is distinct from saying they pick out mental representations. On a Verbial view (Figdor 2014), they pick out activities and modifications of them (e.g. *representing* is a kind of cognitive activity, and *representing that P* is a special case of it and is a mental representation to some). Finally, a Literalist who is an internal-state realist about psychological capacities need not be a brain-state realist, as embodied-mind theorists also deny. Brainless beings are not left out of internal-state realism.

[5] More precisely, the similarity mapping is between elements of the concepts that represent features of the concepts' referents. Below I discuss the problem of what information is contained in concepts of psychological capacities, or, roughly equivalently, the meanings of psychological predicates.

PIGLET, but is used metaphorically in this context to express TODDLER by mapping the two concepts via this shared conceptual element that represents a shared activity of toddlers and piglets.

In parallel fashion, the Metaphor view holds that scientists use psychological terms in the unexpected contexts to refer to a capacity of nonhumans by mapping them to human capacities via similarity comparisons (Gentner et al. 2001: 243 fn. 1). They do this via metaphorical use of a term that conventionally refers to a humans-only capacity. The metaphorical mapping is possible due to similarities between the humans-only capacity and the capacity of the nonhumans. For example, when Hubel and Wiesel (1962) write than neurons prefer certain stimuli, they use "prefers" with the intention of referring to a distinct neuronal capacity that shares common features or relations with real preferring, which only humans have (along with a few other animals). Similarly with this example from Chapter 4:

(2) Resonator neurons prefer inputs having frequencies that resonate with the frequency of their subthreshold oscillations.

On the Metaphor view, this is also a metaphorical use of "prefers". The neurons have something like what humans have when the latter really have preferences (or exhibit them).

Note that the Metaphor view concerns verbal (or relational or predicate) metaphors, not nominal metaphors. Most work on metaphor is on nominal metaphor (Chen et al. 2008). The alleged metaphor in (2) is not a similarity mapping between neurons and humans, but between what neurons do or have and what humans do or have. Scientists are not metaphorically referring to neurons, but are metaphorically referring to a neural capacity by using "prefers", which conventionally refers to a human capacity. The strictly parallel claim for (1) would be that "getting dirty" is an activity of piglets, and because the toddler's activity shares some features with real cases of getting dirty, "getting dirty" can be used metaphorically to pick out their child's activity.[6] Of course, we accept that toddlers and piglets

---

[6] I am not going to fuss about nominalizations of psychological verbs or whether they refer to items classified metaphysically as activities, capacities, properties, states, or processes. Differences in concepts of objects, activities, and properties (or object-, action-, and property-roots: Haspelmath 2012), and how these concepts are manipulated in cognition, are reflected in differences in nominal, verbal, and adjectival polysemy (Gentner 1981; Gentner and Boroditsky 2001; Brown 2008) and brain processing (Damasio and Tranel

alike can really get dirty (and can get really dirty). The Literalist says humans and neurons both really have preferences. The Metaphor view says that neurons don't really prefer anything, but have a capacity that shares some features with real preferring (a cognitivist interpretation) or exhibit behavior that is similar to the preference-exhibiting behavior of humans (a behaviorist interpretation). The Metaphor view is compatible with either a cognitivist or behaviorist interpretation, although for the most part I will assume a cognitivist interpretation.

6.3 The Literalist Responds: Motivation

The first problem is that there is no motivation or evidence for the Metaphor view that is independent of the presumption that the relevant uses just can't be Literal. The contexts of use do not suggest metaphorical intent, standard tests of metaphor don't usher in uniform judgments, and finding other sources of support will be difficult given that the metaphors are verbal. I'll consider each of these subproblems in turn.

Metaphor is often a case of "you know it when you see it".[7] For example:

(3) San Francisco has traditionally been a Dungeness crab of a city, shedding its carapace from time to time and burrowing down until a new shell sets.

(N. Heller, "Bay Watched", *The New Yorker*, Oct. 14, 2013)

(4) Racism is a virus that is growing clever at avoiding detection.

(C. Blow, "Disrespect, Race, and Obama", *The New York Times*, Nov. 15, 2013)

Unlike (1), (3), and (4), it is fair to say that (2) is not *obviously* metaphorical. This isn't trivial, since if it was obvious the Metaphor supporter would no doubt seize upon this fact to dismiss Literalism out of hand. Too bad for me that the fact that these uses are not obviously

---

1993). Figdor (2017) considers some of these noun/verb differences with a focus on psychological verbs.

[7] Shottenkirk (2009: 125), focusing on the metaphysics of possession and exemplification relations on Goodman's extensional view of metaphor, notes that a grey painting is literally grey and metaphorically exemplifies sadness; and we know the former because we look at the extension of "grey" and see if the painting is in it.

metaphorical can't be wielded in equally cursory fashion against the Metaphor view. Nevertheless, further investigation shows that this lack of clear endorsement persists. There's no reason to think scientists intend to use the predicates metaphorically other than if one already believes that Literalism must be false.

Consider background intuitions. Some hold independently that folk psychological attributions to nonhumans need not be metaphorical (Ravenscroft 2010: 3–4). Even if one feels the uses are odd, the oddness can be given different explanations. Searle explained it (see Chapter 5) in terms of importing social aspects, which can be "carved off" to reveal the sensicality of some nonhuman uses. The oddness could stem from a sense of unfamiliarity that reflects non-neuroscientists' lack of knowledge. "Whales are mammals" probably initially sounded odd to many. Nor does historical primacy guarantee semantic primacy. "Computer" originally referred to humans who did a specific job; "computers are machines" probably initially sounded odd to many. Nowadays, uses for humans are metaphorical to some (Boyd 1993) and literal to others (Pylyshyn 1993: 554–6).[8]

Consider the contexts of use. Context often provides clues about metaphorical intent. In the piglet example, the sentence itself is the 1st context of use. The 2nd context is given by the story that surrounds and prompts the utterance of (1): a couple watching their adored child at play. Bezuidenhout's paper is a 3rd context: the sentence and the story are presented in an academic paper in which the author intends (literally) to defend a certain view of metaphorical interpretation. (A 4th context is this book.)

In (2), the 1st context is the sentence in which "prefers" appears. The 2nd is the scientific context that surrounds and prompts the ascription

---

[8] In part this dispute is because "the computer metaphor" does not have one referent. To say "The mind is a computer" (e.g. Bowdle and Gentner 2005: 193) is not the same as saying "Mental capacities are computational". Even in the context of the latter theory, there are warring sides. In response to McDowell's (2010) claim that psychologists' appeals to representation are metaphorical or "as if", Rescorla (2015: 705) argues that representational talk in perceptual psychology "is not heuristic chit-chat". On his view, representational locutions reflect the "explicit goal of the science: to describe how the perceptual system estimates environmental conditions". When Bowdle and Gentner (2005: 193) use "The mind is a computer" as an example of metaphor, they specify that "mind" refers to "an abstract entity" and "computer" to "a complex electronic device". The sentence is metaphorical *because of* these reference assignments.

reported in (2). The 3rd is the peer-reviewed paper in which the sentence appears and the study is reported. (A 4th context is this book.) In these contexts, the main aim of scientists is to express literal truths (or empirically adequate statements). Such contexts do not support Metaphorical interpretation of (2). For example, when a scientist reports:

(5) Ion channels twist open to allow hydrated potassium ions to pass through neuron and cardiac cell membranes (Reuveny 2013),

the sentence is interpreted by default as an assertion of an empirically defeasible truth about cells, ions, twisting open, passing through, and so on. Its constituent terms are used consistently in neuroscience, and its use satisfies Grice's conversational maxims of truth, informativeness, relevance, and clarity (Wilson and Sperber 2008: 250).[9] The contexts of (2) are to all appearances the same as those of (5).

Scientists do use metaphors in these contexts, but they are often marked off in some way, and not just by using scare quotes. For example, Ratcliff (1978: 62) explicitly invokes the metaphor of a tuning fork's resonating to illustrate a possible way to think of the comparison process in decision-making. He distinguishes the drift-diffusion model as such from the resonance metaphor, and uses the latter to gesture towards a way to fill a gap in its explanation of decision-making. However, even this case may best be seen as an epistemic metaphor (discussed further below). It could turn out that resonating and a mechanism that resonates, though not a tuning fork, really are part of the explanation of decision-making. Many metaphorical comparisons in these contexts—as opposed to contexts of popular science or science pedagogy—are of this sort.

Finally, consider standard rules of thumb offered by linguists and psychologists for distinguishing metaphorical from literal use. These rules of thumb are just that: they can help but do not always usher in uniform or definitive judgments of intended metaphorical interpretation. But even with that caveat in mind, the tests fail to provide clear judgments of metaphorical intent in the contexts of interest here. The

---

[9] In Gricean terms, these are maxims of Quality, Quantity, Relation, and Manner. Note that Literalism's spirit of "equal treatment across the board" could be preserved within a nihilist (or otherwise anti-realist) metaphysical framework. A nihilist about (e.g.) ion channels and cells (e.g. Unger 1979; Van Inwagen 1990) would give an account of the truth conditions of (1) in which it is literally true of whatever the nihilist is ontologically committed to. The same would go for (2)–(4).

fundamental problem is that verbs are often used literally across many object domains.

Gibbs' (1990: 57) 'is-like' test is intended to distinguish metaphor (a comparison between two conceptual domains) from metonymy (a part–whole relation in a domain). If a non-literal comparison put in the frame "X is like Y" is meaningful ("The boxer is like a creampuff"), it is metaphorical. If not ("The third baseman is like a glove"), it is metonymy. This test correctly classifies "A toddler is like a piglet" and "Race is like a virus" as metaphorical. But to a Literalist "A neuron's preferring is like a human's preferring" is just a weird way of saying they do the same thing, akin to saying "A dog's seeing is like a human's seeing".

Gleitman et al. (2005: 41) suggest that verbs are "predictably metaphorical" when used in unusual frames. To cite their example, "The reader is sleeping his way through this article".[10] Expanding on this example we get the following:

(6) The reader slept (/is sleeping his way) through the article.
(7) The bear slept (/is sleeping its way) through the winter.
(8) The plant slept (/is sleeping its way) through the winter.

We agree (6) is a metaphorical use of "slept" and that (7) is literal use. What about (8)? Do plants sleep? They have periods of dormancy and circadian rhythms. Does that count? The test doesn't tell us either way.

Now consider these sentences (the last is repeated from above):

(9) Capuchins prefer ready-to-eat fruit to tree bark.
(10) Six-month-old infants prefer attractive faces of diverse types.
(2) Resonator neurons prefer inputs having frequencies that resonate with the frequency of their subthreshold oscillations.

These sentences are all standard in their respective sciences. (9) and (10) are reports of research results. So is (2). But is (2) an unusual sentential frame or an unexpected finding? The Literalist thinks it's the latter. Of course, I am not using Gleitman et al.'s test to show that the uses of "prefers" in all three sentences *are* literal. The problem is that we are not

---

[10] Gleitman et al. (2005) present evidence that credal verbs (*believe, know*) aren't hard to learn because they are more abstract and not clearly associated with observable motions, but because we need to learn the right sentence frames—the right syntax.

obtaining judgments in the Metaphor view's favor that have anything like the clarity of the toddler or racism examples.

Another route to predictable verbal metaphor might be called the troponym test. Verbs often pick out common relations in many nominal metaphors, just as superordinate object categories, such as furniture, often share common functions (Rosch 1973, 1975; Rosch et al. 1976). In (1), a metaphorical relation between "piglet" and "toddler" (or PIGLET and TODDLER) is made via a literal use of the gerund of "gets dirty". Now consider a clear case of verbal metaphor:

(11) The Earth pirouettes around the sun. (Camp 2006)

Ballerinas and planets alike literally rotate. The verbal metaphor involves using a term whose meaning includes reference to a human-specific manner of this common activity. In technical linguistic terms, "pirouette" is a troponym of "rotate".[11] The nominal metaphor in (1), in contrast, employs a verb that is literally true of toddlers and piglets, where species-specific differences in how they get dirty are irrelevant.

This suggests a formula for reliably generating verbal metaphors, and so perhaps for reliably identifying them. Start with a verb used literally across two object domains ("eat", used for humans and pigs). Find a troponym whose meaning includes domain-specific information appropriate to just one of the domains ("dine"). Use the troponym to describe the entity in the other domain: "The pig dined on the garbage". Voilà! Reversing the procedure, if a troponym whose meaning includes domain-specific information is used in a context not specific to that domain, it is likely metaphorical ("He hammered the candidate with tough questions").

Do psychological verbs used in sentences like (2) generate verbal metaphors in this way? It is not clear that they do. It is generally the case that many verbs enjoy a relative lack of object-specificity in their

[11] See Fellbaum (1990) and Fellbaum and Miller (1990) on the troponym relation between verbs ("To V1 (V1-ing) is to V2 (V2-ing) in some particular manner"). The same pattern appears in Carston's (2010: 295) "The river sweated oil and tar" and in the manuscript version of Jacob's (2011) "The ATM swallowed my credit card", where he explains (in the manuscript) that the conditions of applying the concept encoded by the verb "to swallow" are loosened, since swallowing is restricted to living things. (The verbal metaphor is omitted from the published version.) This is in keeping with Chen et al.'s (2008: 200) remark that predicate (verbal) metaphors—at least for verbs of motion—involve omitting or minimizing sensory-motor attributes while keeping more abstract features and highlighting them to establish the metaphor.

literal reference, including ascriptions at multiple levels in part–whole hierarchies (Figdor 2017; see also Chapter 8). You are vastly different from your cells, but it doesn't follow that being alive is only literally true of you. This cross-context appropriateness of many verbs is what makes it easy to formulate so many nominal metaphors. In addition, some psychological verbs are already accepted as being used literally across human and nonhuman domains ("to see"). A Literalist thinks we now have good reasons to think this attitude should generalize to other psychological verbs. In any case, the mere fact that a verb is psychological does not entail that human-specific information is built into the concept it conventionally expresses. We will need an independent reason to think "prefers" is like "pirouette" in the human-specificity of its meaning for the troponym test to be useful. And even if "prefers" were provably human-specific in its meaning, we would not be able to generalize from this particular case to all psychological verbs used in the unexpected domains.

Finally, I've noted that the Metaphor view is less plausible if it restricts literal ascriptions to humans alone. Few agree nowadays that only humans have minds. But by allowing literal uses for some nonhuman species, it must articulate the semantics of psychological predicates in a sufficiently fine-grained manner to show why certain species are within its proper extension while so many others are not. As (9), (10), and (2) show, the scientific contexts of use are too uniform to aid the Metaphor view in this task.

The problem of drawing a distinction between metaphorical and literal ascription comes down to drawing a principled distinction between the information that is part of the meaning of a term and information that is just customarily associated with it. This is the same problem that was raised when considering whether a model construal includes the same or different (in Weisberg's terms, "analogous") assign-ments of relations to a model description's parameters or variables (see Chapter 3). And just as I argued then, the Metaphor view can't simply point to intuitive similarity to humans as the critical meaning criterion for psychological predicates. Doing so begs the crucial epistemic and meta-physical questions behind the semantic debate: what priority should be given to new scientific evidence of similarity across human and nonhu-man domains regarding psychological capacities? Before turning directly to the question of meaning in section 6.4, I'll consider two more tests of intended metaphorical interpretation. The first is from the literature on conceptual metaphor, in which metaphor is a mapping between concepts or conceptual structures that crosses conceptual domains.

Jackendoff and Aaron (1991: 326–7), in a critical review of Lakoff and Turner (1989), suggest an incongruity test for conceptual metaphor, based on the idea that, in standard cases of metaphor, metaphors are considered literally incongruous.[12] The test involves identifying an incongruous cross-domain conceptual mapping and considering the applicability of that mapping to a sentence. For example, is "Our relationship is at a dead end" metaphorical? Suppose the invoked conceptual mapping is A RELATIONSHIP IS A JOURNEY and the sentence is unpacked as follows:

(12) Of course, relationships are not journeys, but if they were, you might say ours would be at a dead end.

The first clause of (12) acknowledges the incongruity of the mapping. The second invokes that mapping; its coherence relies on the conceptual metaphor. This reveals the original sentence as metaphorical.

In contrast, suppose someone says, "My dog ran down the street" on the (dubious) grounds that "run" applies only to people. Suppose the invoked structure is ANIMALS ARE PEOPLE, and the sentence is unpacked as follows:

(13) Of course, dogs are not people, but if they were, you might say my dog ran down the street.

The first clause acknowledges an incongruous mapping, but the second clause is a non sequitur or perhaps a bad pun. One can say that dogs run whether they are thought of as people or not. So the original expression is not metaphorical, at least not if the feature targeted in the incongruous mapping is running.

---

[12] Lakoff and Johnson (1980) argue that human conceptual structure is inevitably structured by metaphor, or cross-domain mappings between concepts (e.g. TIME IS MONEY, ARGUMENT IS WAR). Concepts are categorized depending on whether they arise from direct sensory experience or not. Physical (spatial, concrete) domains yield literal source concepts, while concepts in the non-physical (non-spatial, abstract) target domains are metaphorical. Their view is orthogonal to my concerns because all mental concepts are metaphorical in their proprietary sense of "metaphorical". It is less confusing to think of their view as a version of concept empiricism. For criticism and alternatives, see Jackendoff (1983, 1990); Jackendoff and Aaron (1991); Murphy (1996); Papafragou (1998); Stern (2000); Rakova (2003); Croft and Cruse (2004: 201); Carston (2012: 28).

Now consider the "pirouette" and "piglet" examples ((11) and (1)), using the conceptual structures PLANETS ARE PEOPLE and TODDLERS ARE PIGLETS:

(14) Of course planets aren't people, but if they were, you might say the Earth pirouettes around the sun.

(15) Of course toddlers aren't piglets, but if they were, you might say that our piglet is getting dirty.

(16) Of course, toddlers aren't piglets, but if they were, you could say that our toddler is a piglet (since he, like many a piglet, is getting dirty).

The first clause in (14) acknowledges the incongruity and the second clause invokes that incongruity. Because "pirouettes" is a humans-only manner of rotating, the coherence of saying that the planet pirouettes relies on the incongruent structure. So the original sentence is metaphorical.

In (15), while the incongruity of the structure is acknowledged, the rewrite doesn't capture the fact that the original sentence was a nominal metaphor, not a verbal metaphor. (16) unpacks it as a nominal metaphor. Note that (15) and (16) invoke the same conceptual mapping, but make implicit reference to different aspects. Moreover, if that aspect is not clearly specific to one domain, the test gives the result that the sentence is not metaphorical. Similarly with PLANETS ARE PEOPLE. If we had rewritten Camp's original metaphor as follows:

(17) Of course planets aren't people, but if they were, you might say the Earth rotates around the sun.

the second clause does not invoke a relevant incongruity. The mapping includes rotating as a similar feature, but you don't need to conceptualize the Earth as a person to say that it rotates (as we also saw with the dog running example above). In short, whether the test helps show a sentence is metaphorical depends on whether the relevant aspect of the invoked conceptual mapping is clearly considered to be literally true of only one domain in the mapping (e.g. pirouetting for humans, having dead ends for journeys).

Now let's try the test with a neuron sentence like (2), using the conceptual structure NEURONS ARE PEOPLE:

(16) Of course neurons aren't people, but if they were, you might say neurons prefer inputs having frequencies that resonate with the frequencies of their subthreshold oscillations.

Is "prefers" metaphorical in the original sentence? No doubt intuitions will differ. But the test does help sharpen the debate. Even if the first clause acknowledges the incongruity of thinking of neurons as people, one must *presuppose* that preferring is a humans-only activity in order for the second clause to invoke a relevant incongruity. The Literalist thinks "prefers" is no more restricted to humans than "runs" or "gets dirty". As a result, the Metaphor view is still on the hook for a non-question-begging reason why "prefers" is restricted to humans, as it must be for this test to show that (2) is metaphorical.

A final attempt to distinguish metaphorical from literal use is suggested by Pylyshyn (1993: 455) in the course of defending his claim that computational language for describing the mind is literal. Pylyshyn provides two pragmatic desiderata for distinguishing metaphorical from literal use in science: (a) the stability, referential specificity, and general acceptance of terms, and (b) the perception shared by those who use the terms that the resulting description characterizes the world as it really is, rather than being a convenient way of talking about it or a way of capturing a superficial resemblance. These desiderata are to the point in that they target scientific usage, but they are too weak to settle the debate.

First, our judgment regarding (a) depends in part on what evidence is considered relevant for determining reference. "Prefers" in neuroscience already satisfies (a) in that its uses in neuroscience are stable and generally accepted. But the question of referential specificity depends on whether one includes the evidence from modeling practices as a basis for referential stability and what epistemic priority it is given. If one takes for granted the traditional anthropocentric semantics for psychological predicates, "prefers" in neuroscience cannot satisfy (a) in a way consistent with Literalism. But this begs the question against Literalism. Second, the open question in (b) is what the resulting description is about. We want to characterize the world as it really is. But what is that? Literalism holds that mathematical models of cognition are providing new evidence of what psychological capacities are really like that no longer depends on our anthropocentric lights. Nor is this the only possibility. Technical views, discussed in Chapter 7, are also consistent with (b).

As a final point relating to the issue of metaphorical intent, I noted above that if the neuron sentences strike one as odd, this could be due to a lack of empirical knowledge. This is consistent with the Technical view, in which "prefers" has a different referent in these contexts and laypeople lack the knowledge to recognize the difference (see Chapter 7). In contrast, the Metaphor view must hold that scientists share the lay adult's understanding of "prefers" when they use the terms in these contexts. "Prefers" means what laypeople think it does, and the scientists are using it with metaphorical intent.

So it is informative to consider this claim in the light of the conceptual transition from child to adult concepts (e.g. Carey 1988; Barrett et al. 2001). This transition occurs between ages 4 and 10, after most children demonstrate understanding of the concept of belief and while they are still coming to understand metaphor (Winner et al. 1976; Vosniadou et al. 1984; Inagaki and Hatano 1987, 1993, 1996; Winer et al. 2001). Children that pass the false belief task—around age 4—have a concept of belief that is not quite the adult concept (Carey 1988). At a later age, the child learns that certain kinds of things (e.g. puppets) can't ever really have beliefs, not even by magic, because she learns that there is no such thing as magic. The concepts of the "biologically unsophisticated adult" (Carey 1988: 176) are better than the child's in part because the adult is sensitive to empirical facts the child has not yet learned, and the adult's concepts reflect this knowledge.

The Literalist holds that the biologically sophisticated adult is sensitive to more and newer empirical facts about psychological capacities, and her concepts reflect this knowledge. In contrast, the Metaphor view entails that the correct understanding of psychological concepts is the biologically unsophisticated adult's. But why should our correct understanding of the psychological stop there?

6.4 The Literalist Responds: Meaning

The second response is that standard ways of unpacking the Metaphor view's semantics for psychological predicates do not show that their meanings rule out literal use in the unexpected domains. The underlying issue is to determine what information is contained in psychological concepts. The Metaphor view holds that the predicates are being used metaphorically in the relevant scientific contexts to pick out meanings other than their conventional meanings, via similarity relations. But what are their conventional meanings? The Nonsense view argued explicitly that these uses violate semantic conventions. The Metaphor view in effect agrees, but avoids saying so explicitly by arguing that the uses are not intended to invoke their conventional meanings. In the last section, we found no independent sign of such intent. In this section, we'll look at the Metaphor view and conventional meaning from the perspective of two standard general semantic theories—classical semantics and relevance-theoretic semantics—and their respective explanations of metaphor.

The classical or indirect view of metaphor associated with Grice (1975/ 1989) starts from the idea that the meaning of a non-indexical term is semantically (lexically) encoded in the concept it expresses. A use of the term contributes this same context-independent meaning to the proposition expressed by any sentence in which it appears (see also Davidson 1979/2001). This context-independent or literal encoded meaning fixes literal use. Metaphor arises when hearers (or readers) recognize the blatant flouting of the Gricean conversational maxim of truth-telling and infer to a distinct proposition than the one encoded in the uttered (or written) sentence.

For example, if Eugene M. Izhikevich were to declare "My job is a jail!", this sentence encodes the proposition (fixed in context by assigning Izhikevich as the referent of the indexical "my") *that EMI's job is a jail*. The obvious falsity of a literal interpretation prompts us to infer to the proposition *that EMI's job is confining* (or something similar). So on this account, when Izhikevich writes:

(2) Resonator neurons prefer inputs having frequencies that resonate with the frequency of their subthreshold oscillations,

the Gricean account claims that readers see the obvious falsity of the sentence and infer to a distinct proposition than the one literally encoded in the sentence. Izhikevich uses (2) metaphorically to express an alternative proposition about a distinct capacity of neurons that has common features with real preferring.

The Gricean account does not do the Metaphor view any favors, because (2) strikes neuroscientists as true (or empirically confirmed), perhaps even obviously true. It certainly does not strike them as obviously false. So there is no reason for them to infer to another proposition than the one that seems to be semantically encoded in the sentence. There is, of course, this alternative proposition:

(2a) Resonator neurons respond selectively to inputs having frequencies that resonate with the frequency of their subthreshold oscillations, which was ISQ #2 (see Chapter 4). In my response then, I pointed out that one could just as easily substitute "responds selectively to" for "prefers" in human sentences. The debate then turned to clarifying the difference between them. For present purposes, whether (2a) is an adequate literal paraphrase of what the neuroscientist is expressing metaphorically by using (2) is not the issue. The issue is that the view presupposes that "prefers" expresses (or PREFERS contains) human-specific information that rules out literal use for nonhumans. The motiv-ation to interpret (2) metaphorically *follows from* this prior semantic commitment. We've advanced matters in that the Metaphor view's discomfort with (2) has been articulated more precisely, but we still lack a reason to accept the prior semantic commitment. We still need to know what is in the context-independent meaning of the relevant psychological predicates and what isn't.

In any case, the classical idea that we initiate metaphorical processing after literal processing fails is no longer the received view of metaphor comprehension (Gentner et al. 2001: 236; Gibbs and Tendahl 2006). So the Metaphor view may be better off trying to elaborate the meanings of psychological predicates within the context of the current main alternative theory, relevance-theoretic or pragmatic semantics. RT, as I will refer to it, rejects the classical semantic view of meaning as largely or wholly context-independent (Sperber and Wilson 1986; Recanati 1993, 2010; Bezuidenh-out 2001, 2002; Carston 2002, 2012; Wilson and Carston 2007; Wilson and Sperber 2008).[13] On this view, the context-dependent concept expressed by "prefers" in (2) is semantically related to the concept it expresses when "prefers" is used for humans. This is helpful to the Metaphor view because it plausibly allows that the unexpected uses retain semantic ties with literal uses but still count as metaphorical in the new contexts.

The core idea of the RT approach is that all lexical terms in natural language are subject to a great deal of contextual variation. For example,

---

[13] Classical semantics allows two basic sorts of context-dependent meaning adjustments: saturation (when an indexical, such as "I", is assigned a referent) and disambiguation (when one meaning of an ambiguous term, such as "bank", is selected). RT holds that these two adjustments are not sufficient to yield the complete meaning or proposition that hearers pick out. Classicists may also argue that terms (in sentences) contain hidden variables in their logical form, while RT theorists say the concept is augmented with unarticulated constituents provided by context that are *not* encoded in the logical form of the sentence. These and other debates about which proposition a sentence expresses and which infor-mation expressed by a term is stored in memory do not matter in this context.

the word "cut" has subtly different meanings in "She cut the cake", "She cut herself", and "She cuts herself". RT argues that the context-independent (or lexical or encoded) meaning that a term contributes to the content of a token sentence in which it appears is modulated using information provided by the sentential and communicative context.[14] Lexical or encoded meaning includes encyclopedic and logical information (Wilson and Carston 2007: 29–31). Adjustments to this lexicalized core involves such processes as dropping encyclopedic information (e.g. PRINCESS* drops being of royal birth when characterized of an imperious woman). Other times it involves category-crossing (e.g. BULLDOZER* said of an obstinate or persistent person). Such modulation is variously called ad hoc concept construction, contextual (or free) enrichment and extension (or loosening), or pragmatic adjustment of conceptual encodings or of lexical or encoded meaning (Glucksberg and Keysar 1993; Bezuidenhout 2001; Carston 2012).[15]

These semantic adjustments in context yield a continuum from literal use to loose use (or loose talk) to metaphorical uses, where loose use is still literal. Metaphorical interpretation is a radical form of context-dependent adjustment. For example, suppose FLAT is the concept of flatness in geometry. When Omaha-based investor Warren Buffett says "Nebraska is flat", his use of "flat" expresses an adjusted concept FLAT* that is appropriate to terrestrial geography. (For some, the concept is the same, but this will not matter here.) In this loose use, it's literally true that Nebraska is flat, but "flat" in Buffett's sentence does not mean exactly what it does in "This geometric surface is flat". But if Buffett says, "Nebraska is a pancake", "pancake" expresses PANCAKE**. This ad hoc

---

[14] The context-independent core is also called what is linguistically encoded by the meaning of a word (Jacob 2011: 26), its lexical or literal encoded meaning (Carston 2012 fn. 15), its t-literal meaning (Bezuidenhout 2001; Recanati 2001), or linguistically encoded concepts (Wilson and Carston 2007; Carston 2012).

[15] Because of this context-dependence of much of meaning, truth conditions (or propositions expressed by sentence tokens) are far more fine-grained for RT than for classical semantics. It follows that the contributions to truth conditions of "prefers" in human and neuron sentences are bound to differ, but this is just because they almost always differ between any two contexts, even if we are only talking about humans. RT also makes sense of Millstein et al.'s (2009: 8) claim that the concept of drift unites a number of different models of drift in evolutionary biology. This can only happen if the concept DRIFT can sustain somewhat different interpretations for different models yet provide a common core interpretation for all of them.

concept is a radical modification or adjustment of PANCAKE, which encapsulates information about the food in its encoded meaning. The radicalness of the adjustment justifies classifying Buffett's "pancake" sentence as metaphorical, as compared to a literal use when he utters "I ate pancakes for breakfast".

RT thus enables the Metaphor view to account nicely for the sense that the concepts picked out by "prefers" in neuron, capuchin, and human sentences are related and may be literal in some nonhuman contexts without having to accept that it is literal for neurons. The idea is that the lexical core of PREFERS includes information that if left out results in an ad hoc concept that is radically different from the standard concept, and that this information is left out in neuron sentences to yield metaphor. "Prefers" in (2) expresses an ad hoc concept PREFERS** that is radically modified from the core concept PREFERS, while "prefers" in the capuchin and infant sentences (9) and (10) express either PREFER or the loose but still literal PREFERS*.

The RT framework does provide a way to articulate the Metaphor view in more detail. However, it does not establish the Metaphor view unless one already assumes that Literalism is false. I'll show this by using discussions of meaning adjustment in RT that are independent of this debate. It's a bonus for me that they also provide additional support for Literalism, in particular its Anti-Exceptionalist component.

Clear cases of information that is considered encyclopedic for concepts about natural phenomena often depend on scientific discoveries and their precise elaboration using appropriate quantitative methods. For example, Carston's (2010: 303; 2012: 479) discussions of meaning adjustments employ the concept BOILING, the lexical or encoded core of which is determined in part by the physical facts about water. "Boiling" may express BOILING (the scientifically determined boiling point) or BOILING* (a temperature range hot enough to be painful). Both share HOT as elements, but we can still say literally that the water in the teapot is boiling even if it has not quite reached the scientifically determined reference point. The same might be said of the concept GOLD. "Gold" may be used literally (as loose use) to describe an Olympic medal even if it is not pure gold, given the scientifically determined reference point of GOLD.

Note that the idea of a context-independent core does not imply that conceptual revision entails fully distinct concepts before and after the adoption of scientifically established reference points. There was no such implication for BOILING or GOLD—rather, both concepts illustrate continuity of referential intent before and after science did its job.[16] It also does not imply conceptual determinacy or stagnation or the impossibility of revision over time. In cognitive linguistics, pragmatic semantics, and related fields, the core is described in dynamical terms of relative semantic stability (Barsalou 1987; Carston 2002; Gibbs and Tendahl 2006: 395). In philosophy, Wilson (2006: 391) argues that concepts are associated with "directivities" ("standards of fairly trustworthy measurement and inference") rather than denotations; predicates "carry with them relatively permanent bundles of directivities which are open to our inspection and modification" (2006: 99).[17]

Without reference points, intuitions about what is encyclopedic can be variable and unreliable. Judgments of the centrality or typicality of exemplars of a category are unstable within and between categories (Rosch 1975; Barsalou 1987). Lakoff and Turner (1989: 57, 134, 194) hold that we metaphorically ascribe human characteristics, such as loyalty, to dogs, which (they assert) only act out of instinct. On their view, there is a moral component to genuine loyalty. In contrast, Jackendoff and Aaron (1991: 331) argue that LOYALTY has as a major component WILLINGLY STAYS WITH X WHEN X IS IN TROUBLE and that this characterizes dogs and people equally. On their view, even if human loyalty includes a moral dimension, "loyal" can be contextually modified to leave out a human-specific moral dimension and reference to a dog's loyalty would be literal.

Literalism holds that psychological concepts are acquiring scientifically determined reference points that can ground a context-independent core of PREFERS that is not human-centered or human-specific. Mathematical models, such as those discussed in Chapter 3, promise to provide these reference points. There is no guarantee that scientific discovery will

---

[16] See also Chang (2012) on the concept of acidity and its relation to lay meaning.

[17] Wilson is also a scientific realist (2006: 84). On Wilson's view, the "classical" picture errs in holding that the core is like a rule that determines beforehand the adjustments we will need to make as we confront the world. Instead, we "possess real but limited control over the wanderings of our words and should not unwisely demand more" (2006: 104). I agree, but since Wilson's target is classical semantics in philosophy of language and mind, his emphasis (in contrast to mine) is on the slippage rather than the stabilities.

map the nonhuman entities apt for literal ascription in intuitive order on the continuum of semantic adjustment.

In the current absence of such a reference point for "prefers", however, suppose we agree that RESPONDS SELECTIVELY is contained in the lexical or encoded meaning of "prefers" across human and neuron sentences. The Literalist can say that this core concept PREFERS can be adjusted to accommodate different types and ranges of stimuli, processing methods, and levels of predictability. Emotional accompaniments can be contextually included in some but not all human uses. In addition to strictly literal use, there can be many ad hoc concepts PREFERS* that are literal but loose in both neuron and human contexts. Referential continuity across human and neuron domains also doesn't entail that all uses for neurons are loose and all those for humans are precise (or vice versa), since looseness won't be determined by an anthropocentric semantic standard. Nor is the Literalist required to hold that "prefers" is always used with referential precision. Even when we have a precise concept (e.g. BOILING), we can prefer a looser concept (e.g. BOIL-ING*) in context. For example, we can say of boiling water that it is BOILING* intending to be literal but not pedantic. All of these possible differences in meaning in context are compatible with Literalism.

The Metaphor view supporter might respond that by leaving out the emotional dimension or other human characteristics we really have gutted the concept. It no longer contains a core element. The Literalist view is akin to claiming that "Nebraska is a pancake" contains a literal use of "pancake" even though the ad hoc concept it expresses leaves out CONTAINS FLOUR.

But this response just asserts what it in dispute. Without an argument for specific additional encoded meaning components (besides RESPONDS SELECTIVELY) that rule out literal use for neurons, we still don't have a non-question-begging reason to think the neuron sentences are metaphor-ical. The objection appears to reject the very idea that PREFERS is subject to scientifically determined reference revision that would undermine the anthropocentric semantic standard for psychological terms.

I conclude that the Metaphor view fails as a plausible alternative to Literalism. In a nutshell, it asserts what it needs to show. It is not backed by clear independent motivation for or evidence of metaphorical intent, and it does not show that psychological concepts contain human-specific information in their context-independent core.

6.5 Epistemic Metaphor

The other sense of "metaphor" that may be involved in claims of metaphor is what might be called epistemic metaphor. This sense does not threaten a Literalist interpretation, so I will discuss it only briefly.

Epistemic metaphor is analogy, and analogy is integral to our efforts to accommodate our theories, concepts, and language to the causal structure of the world (e.g. Hesse 1966). A paradigm of analogical comparison is the Bohr model of the atom, which proposed that electrons orbit around a tiny dense nucleus the way planets orbit the sun. Another example is the Hodgkin–Huxley model of the neuronal action potential, which relied "on a metaphor: the equivalent electrical circuit. By extending this metaphor beyond passive membranes, it captured vast amounts of data and guided decades of research into the underlying biological hardware (voltage sensitive ion channels)" (Carandini 2012: 508–9). Note the intended literal use of the verb inside this nominal metaphor: ion channels are not really electrical circuits (understood as having wires, switches, transistors, etc.) but in common with real electrical circuits they are really voltage sensitive. In other analogies, this sameness of activity is all that matters. Reuveny (2013: 183) likens the conformational changes of an ion channel as it opens to the conformational changes of a lens aperture of a camera when it is hand-rotated open. The ion channel is not called a lens aperture; Reuveny does not offer a nominal metaphor or analogical model of the Bohr type. Instead, he reports what is going on literally at a less observable scale by using a term that also refers literally to more familiar cases of the same process.

None of this is controversial or new. The important point is that it is compatible with Literalism. Epistemic metaphor involves making a metaphysical comparison for epistemic purposes of proposing and communicating new or otherwise undeveloped ideas or theories. This is different from using an analogy or metaphor to communicate science to laypeople or students. In science, when an old term is introduced or justified via analogy, the term often retains enough of its old meaning to guide further theorizing while being open to revision in the light of further investigation. Sometimes the meanings are the same, as in the "orbits" and "twists to open" examples. They are used literally with no change in reference across contexts. In other cases, the terms are informative placeholders that indicate explanatory or evidential gaps.[18] In these cases, epistemic metaphor involves uses of terms in new contexts that indicate regions of semantic space whose contours are in flux. The terms are "usefully vague" (Godfrey-Smith 2005: 4; see Chapter 3 for further discussion). This is compatible with a Literalist semantic outcome for the extended terms. If a term that is introduced to help fix reference does help to fix it, we have a good reason to keep it once reference is fixed. In the case of psychological predicates, we're just now acquiring evidence that supports extending the terms to the new contexts. The prudent scientific option is to refrain from declaring meanings fixed prematurely. This is consistent with the Literalist's idea that the reference of these terms is being revised in the light of the new evidence.

Epistemic metaphors are often defended on grounds of theoretical utility. The qualitative analogical extensions of psychological predicates in the cases of plants and bacteria discussed in Chapter 2 are often defended this way. I noted that sometimes these uses are called metaphorical, sometimes literal. This lack of uniformity among scientists does not show that Literalism is imposing an interpretation on them, because these claims need not be contradictory. The claim that the extended uses are non-metaphorical semantically is compatible with the claim that they are metaphorical epistemically.

6.6 Concluding Remarks

I have argued that the Metaphor view does not have independent motivation or evidence in its favor, and that it asserts but does not show that psychological terms refer to human-specific capacities. I tried out a number of tests that are intended to distinguish metaphorical from literal use. The Metaphor view failed them all. I also considered two general theories of semantics and their accounts of metaphor to see if they might help. They provide more theoretical machinery for discussing meaning, but they don't provide independent evidence for the Metaphor view. Finally, epistemic metaphor is a claim about analogical extension in science, and is consistent with Literalism.

---

[18] Because uses of psychological terms in nonhuman domains provoke such strong responses, they are highly unlikely to be anodyne "filler terms" (Craver 2007), like "cause" or "make". Filler terms are useful indicators, but they don't indicate much.

Up to this point, I have demonstrated at the very least that it is a mistake to think Literalism is so obviously implausible that actual argument against it is otiose. When we begin to consider alternatives in detail, the Nonsense and Metaphor views turn out to be implausible or unsupported. However, the anti-Literalist options are not yet exhausted. If the uses of psychological predicates in the unexpected domains are neither non-sensical nor metaphorical, then they are in some sense literal. But they may not be Literal; they may be literal in some other sense. The term "technical" is sometimes used to signal a claim that they are used sensically and literally, but with different reference. I adopt this term to label the remaining group of alternatives, presented in Chapter 7.

7

The Technical View

7.1 General Remarks

In this chapter I consider a final alternative to Literalism. As the label indicates, the Technical view holds that the uses are technical in some sense. This is a technical sense of "technical", since merely saying a term is technical is not in itself anti-Literalist. Technical definitions are often considered the standard for literal truth both within and outside of science. They may indicate a term's scientifically determined reference. The Technical view, however, asserts a difference in reference between human and nonhuman uses of psychological predicates. It is a way of claiming that old and new terms have gone their separate referential ways, even if they stay in touch from time to time. Other ways to indicate this view are that the uses are "shorthand", "a gloss", "deflated", "attenuated", or involve Dennett's Intentional Stance.

Like the Nonsense and Metaphor views, the Technical view is a conservative strategy for interpreting psychological predicates. It denies the Anti-Exceptionalist metaphysics of Literalism. Like all interpretations other than the Nonsense view, it holds that the terms are used in the nonhuman domains with sense. Like the Metaphor view, it holds that the terms are used in these contexts to refer to something other than humans-only capacities. But unlike the Metaphor view, it holds that in the new contexts they are used directly to refer to distinct referents. It maintains the human-centered standard for the reference of psychological terms and so entails that this reference is not being adjusted in the light of scientific advances. Instead, we are discovering distinct new properties or capacities of nonhumans. Literalism, in contrast, holds that the reference in the new domains is not distinct because the standard itself is being adjusted in the light of new evidence to include human and nonhuman domains alike.

As with the Metaphor view, my presentation of the Technical view is more detailed than I have encountered in print or conversation. I divide the Technical view into two basic variants: the Technical-Behaviorist variant and the Exsanguinated Property variant. My elaborations rely on familiar ideas from other debates that have been alluded to in this context as well. Different objectors may be noncommittal between one variant or another. Dennett in particular can be interpreted in various publications as suggesting both Technical variants as well as a Metaphor view, although I think his deepest commitment is to the Exsanguinated Property variant. However, it is not critical to my discussion which view he or any other theorist actually supports, since my target is the view itself.

The Technical view variants differ fundamentally at the metaphysical level, regarding what each claims the predicates refer to when used for nonhumans. On the first variant, the terms refer to patterns of behavior. I assimilate the Intentional Stance into this variant for reasons to be explained below. On the second variant, they refer to properties or capacities of nonhumans that are not fully cognitive. I assume either variant could be elaborated more fully within a standard semantic theory. I also consider them more plausible than the Metaphor and Nonsense alternatives because they imply a positive correlation between more knowledge and more literal uses. Our language is accommodating to the new knowledge, just not in the way the Literalist holds. However, I will argue that both variants fail to provide independent, non-ad hoc justification for thinking that their referential distinctions will align with distinctions in ascriptions to humans and nonhumans. Any other variant of the view would face the same challenge.

As noted, the Technical view's core claim is that there is a referential distinction between what the psychological terms (or word forms) refer to when used for humans and when used for nonhumans. This is not contextual variation of meaning, which is consistent with sameness of reference and Literalism. The question is whether they are literal with the same reference (Literal) or literal with distinct reference (Technical).

It follows that, on this alternative, psychological word forms pick out homonyms or distant polysemes, both of which are forms of ambiguity. A paradigm example of homonymy is "bank":

(1) The *bank* was slippery from all the rain.
(2) The *bank* was closed for the holidays.

These are two distinct senses, each with its own referent, of "bank", with a sharp meaning boundary between them. A strong Technical view (on either variant) holds that the terms in human and nonhuman uses are homonyms, akin to $bank_{financial}$ and $bank_{geological}$. A common heuristic for a sharp meaning boundary is the zeugma test (Croft and Cruse 2004: 113). For example:

(3) John and his driver's license expired last Thursday.

In this case the homonymous terms ("to expire") have distinct meanings that create a sense of pun when the word forms are used in a single sentence. ("The bank was slippery and open until 5" might do the job for "bank".) If there is no such sense of pun, the terms are not separated by a sharp meaning boundary. We might use the zeugma test to check our semantic intuitions regarding psychological word forms:

(4) Infants and neurons *habituated* to the novel stimuli.
(5) Human subjects and neurons *preferred* specific stimuli.

I include (4) for comparison, using another term that appears in scientific contexts for humans and neurons. Of course, both sentences are highly artificial. But by this heuristic there does not appear to be a sharp meaning boundary that favors a claim of homonymy for either italicized term. Neither is clearly a pun.

A weak Technical view (on either variant) holds that the asserted referential difference grounds distant polysemy, rather than homonymy. A paradigm example of a polyseme is "run", which has a large number of polysemes—e.g. "He *runs* the shop" and "He *runs* the track" (Brown 2008: 5). Articulating the semantic differences between polysemes is difficult. But because the Technical view has more significant problems, I set this issue aside. All the view requires is that the predicates are sufficiently distinct in reference in the new contexts from their reference in human contexts, where they may be polysemous anyway (Brown 2008; WordNet <http://wordnet.princeton.edu>). Thus, as I elaborate the Technical view, it holds that the new terms are distant polysemes that can nevertheless retain important semantic links with their human-appropriate cousins.

7.2 The First Variant: The Technical-Behaviorist

With some significant massaging, the most familiar Technical-Behaviorist option may be the Intentional Stance (Dennett 1978b, 1981/1997, 1987). For example, Van Duijn et al. (2006: 166) contrast characterizing the cognitive capacities of bacteria as "genuine decision making" with taking the "intentional stance" towards them, without endorsing either option. The claim is that when we use these terms for humans we are ascribing genuine psychological states, but when we use the same word forms for nonhumans we are merely taking the Intentional Stance towards them. This is not Dennett's understanding of the Intentional Stance, but that label has been used for this variant anyway. Hence the significant massaging.

The Intentional Stance holds that there is nothing more to belief and other mental states than taking a particular kind of fruitful explanatory perspective towards an object, whether that object is a lectern, a human, a neuron, or anything else. It is often considered a form of instrumentalism or anti-realism about the mental, although Dennett (1991) also defends a quasi-realist metaphysics.[1] One presumes practical rationality on the part of the target system—the systematic interplay of the explanatory states that promotes its adaptiveness—and ascribes to that system states with contents that maximize the practical rationality of its behavior. Understood this way, it is a general interpretation of psychological predicates that contrasts with a more traditional cognitivist view in which the predicates pick out mental states, typically internal states, that do not depend on a third party's assuming a particular perspective towards the system that has them. While the Intentional Stance is invariably presented in terms of ascriptions of belief, desire, and intention, it applies to psychological states in general. For example, "expect" and "decide" are sometimes considered explicitly (Dennett 2009). These terms, unlike "belief" or "desire", also figure among the unexpectedly broad ascriptions in biology.

---

[1] It doesn't matter here whether the Intentional Stance is one view across time or whether "the Intentional Stance" is ambiguous between anti-realist and quasi-realist versions. Nor will it matter whether every Intentional system must be able to take the Intentional Stance towards itself or other systems.

Because Literalism could be articulated consistently with instrumentalism or anti-realism, the Intentional Stance is not an *alternative* to Literalism unless its adoption entails a *distinction* between psychological predications to humans and those to nonhumans. Dennett certainly doesn't understand it that way. On his view all ascriptions are (or are the result of) stances, of which the Intentional Stance is just one (albeit the critical one for explaining certain phenomena). The Literalist has a similar non-discrimination policy. So to be a Technical variant, the view must be that psychological ascriptions to nonhumans involve the Intentional Stance and those to humans do not. One might also put the view by saying that the referents of psychological predicates are real or genuine when used for humans but not nonhumans. However, putting it this way risks losing sight of the fact that the Technical view holds that the terms are used literally (but not Literally) for nonhumans. They simply refer to something else, which is frequently described as not genuine or real compared to the human reference. The label "Intentional Stance" is adopted here with this non-Dennettian difference in mind.

So with this caveat, the important question is this: what is this difference in what is denoted when one adopts the Intentional Stance as opposed to when one doesn't? One might articulate the difference by saying that the terms refer to real cognitive states when used for humans, but to abstract entities akin to centers of gravity or other useful fictions when used for nonhumans (Dennett 1991).[2] This is a possible metaphysics, but I hesitate to pin it on a Technical view supporter. I've had interlocutors suggest a behavioristic reading of the uses (see ISQ #2 in Chapter 4), but never an abstract or fictional entity account. Also, while fictionalism about models is a live option (e.g. Frigg 2010), it is a general view that does not metaphysically distinguish models of humans and nonhumans.

---

[2] As Dennett (1991) notes, some think centers of gravity are real and others are anti-realists about them. The important issue here is the referential difference being drawn, whatever one counts as real. The "real patterns" view of the Intentional Stance (Dennett 1991; Haugeland 1993) holds that a pattern in some data set exists or is real if the data can be described in a way that is more efficient than just describing every element in it. For example, one can describe an arrangement of dots (with a temporal dimension) as a glider in the Game of Life computer simulation. In these terms, construals of mathematical model descriptions (equations) describe real patterns of behavior. The Literalist emphasizes that these patterns are discerned from a non-anthropocentric perspective, with the consequent implications for psychological predicates in the case of cognitive models. The Technical view would still hold that cognitive models differ in reference when the target systems are nonhuman.

So it seems more plausible to hold that when we adopt the Intentional Stance for nonhumans the intended reference is in some sense behavioristic. When we adopt the stance, we use the predicates to denote patterns of behavior that are useful for prediction and explanation. This yields the Technical-Behaviorist view.[3] For example, when neuroscientists use "prefers" for neurons they are taking the Intentional Stance towards them, which entails ascribing to them a response profile and nothing more. With humans, we don't adopt the Intentional Stance and so use the terms to ascribe cognitive states that do not depend on a third party's perspective to be possessed. Neural response profiles may be similar to certain human behavioral patterns, but the similarity merely prompts extension of the word form "prefers", not the word itself. The Technical-Behaviorist might even cite these behavioral similarities to explain why "prefers$_{nonhuman}$" and "prefers$_{human}$" are polysemes, rather than homonyms, despite the referential difference.

It follows from this view that neither quantitative nor qualitative analogy justifies an inference to non-perspectival psychological states in nonhumans. (They need not be internal states, but I'll assume so for ease of discussion.) In the case of the model-based extensions discussed in Chapter 3, the cognitive models used for humans and nonhumans are models of internal cognitive states only when used for humans. For example, one of DasGupta et al.'s (2014) explanations of the mutant fruit flies' significantly worse response times in low-contrast stimulus conditions was that the FoxP mutation affected their ability to accumulate the evidence needed to make a decision. For the Technical-Behaviorist, the use of "decides" in this context entails that they are taking the Intentional Stance towards the flies and therefore ascribing the alternative reference. Disambiguated, the flies decide$_{technical-behaviorist}$ but don't decide$_{cognitive}$. Similarly, neurons anticipate$_{technical-behaviorist}$ but don't anticipate$_{cognitive}$, or they prefer$_{technical-behaviorist}$ but don't prefer$_{cognitive}$. We may also express this, albeit less perspicuously, by saying that neural and fruit fly states are not genuine or real, in contrast to the human states, which are.

---

[3] Why not just be a behaviorist about ascriptions to nonhumans without the detour through the Intentional Stance? Fine with me. But explaining it this way reveals why the Intentional Stance gets invoked as a response to Literalism despite its official neutrality between human and nonhuman ascriptions. I discuss behaviorism further below.

7.3 The Literalist Responds to the First Variant

How do we know when the Intentional Stance is being adopted and the reference of psychological predicates (or word forms) has changed? The Technical-Behaviorist responds: when we're talking about nonhumans, we adopt the Intentional Stance, and when we're talking about humans we don't. Really, though, if she is going to beg the question, she should try being a bit more subtle. So let's consider why this mapping of Intentional Stance ascriptions to nonhumans, but not humans, is untenable.

Consider model-based extensions, such as those presented in Chapter 3. In effect, the Technical-Behaviorist asserts that the New Argument for other minds is not cogent without saying what blocks its cogency in the move from the Old Argument to the New one.[4] Without independent evidence that shows the terms are being used in the different ways that the Technical-Behaviorist requires, her distinction is an ad hoc imposition on a pervasive explanatory practice in science. The objection is not that model construals are never modified. It is that our intuitions about what feels like it's the same construal and what feels like it's different are no longer in the driver's seat. A mathematical model provides strong evidence that two domains have important similarities whether or not intuition agrees. Mathematical models of cognitive processes are not in a class of their own in this regard. But the Technical-Behaviorist must insist that they are, even over the explicit intentions of modelers such as Sutton and Barto (1981) to develop models that are appropriate

---

[4] For ease of reference I repeat them here from Chapter 4:

(OA)
1. My behavior is caused by my mental states.
2. I observe that others behave similarly to me.
3. Either they have mental states that cause their behavior, or I am unique and something else causes their behavior.
4. The first hypothesis is best because it explains both cases.
5. So it is probable (or rational to conclude) that they also have mental states.

(NA)
1. My behavior is caused by my mental states.
2. Scientists have developed quantitative cognitive models of my and others' behavior.
3. Either the models' construals are the same for others or I am unique and we must interpret the construals differently for the others.
4. The first hypothesis is best because it explains both (or all) cases.
5. So it is probable (or rational to conclude) that they also have mental states.

for a wide range of adaptive systems. So the Technical-Behaviorist must motivate her view in the face of these apparently non-discriminatory practices. She must defend a non-question-begging principled distinction between model construals that will yield her desired partitioning of the two kinds of ascriptions (Intentional Stance-y or not) between the two classes of target entities (nonhuman or human).

With respect to qualitative analogy, I noted in Chapter 2 that one *can* always interpret qualitative analogical uses in purely behavioristic terms. In these contexts, a Technical-Behaviorist type of response—if not one that invokes the Intentional Stance—is already familiar from and looms large in debates over animal cognition. By many interpretations of Morgan's Canon (but not all: Sober 2005), the canon directs us to get by with a behavioristic explanans whenever we can. In this way the burden of proof is put on those who think cognitivist explanations of animal behavior are justified. But the epistemic upshot of the new discoveries displayed in Chapters 2 and 3 is that the burden of proof has shifted. The behaviorist must now defend her key background assumptions that (i) nonhumans are too simple or inflexible to be ascribed genuine cognitive states (see Chapter 2), and that (ii) what makes a state "genuine" is determined by and relative to the intuitively understood human case (see Chapter 3). If these assumptions do not also motivate the Technical-Behaviorist, we still need a motivation for her referential distinction and a principle that will entail the mapping of behavioristic ascriptions to nonhumans and cognitive ascriptions to humans.

ISQ #3 (in Chapter 4) already attempted to partition humans from nonhumans based on flexibility or variability in behavior. What might be added now to that earlier discussion is overt recognition of the anthropocentric idea that only humans are unique individuals and not mere members of a kind or class. When we are deprived of our individuality, it is felt as a fundamental affront to one's humanity. No one wants to be a clone. So the Technical-Behaviorist might say that when we ascribe real preferring, we imply that the entity can behave against type and thus affirm her status as a unique individual (in the sense in which "unique individual" is non-redundant). Her uniqueness depends on her potential for variability. (I think this is a very bad move, but I won't discuss it until Chapter 9.) So nonhumans do not really prefer because they do not have the capacity to behave in ways that make them unique. They may have some variability, perhaps even more than we thought, but not enough to count as unique individuals. We ascribe Intentional Stance-y states to such non-unique individuals.

It's true that the idea that nonhumans are unique individuals is barely thinkable except in the case of pets or hand-raised farm animals (and occasionally some lab animals and special trees). This attitude is reinforced by the fact that in many cases we only observe a tiny subset of the nonhuman populations, often in artificial circumstances and with explanatory purposes that enforce internal and behavioral uniformity by design. But this very deep form of anthropocentrism is mistaken. We may not care to investigate neurons as individuals most of the time, and we certainly don't care about them in any other non-instrumental sense. But neurons, like other individual living nonhumans, are unique individuals. Rolls et al. (2005: 115 Fig. 1) lists a few of them, named in a way considered dehumanizing for humans. Take be068. Maybe be068's configuration of dendrites and synaptic connections make it differ in its responses in subtle ways that we aren't yet able to distinguish. If its response profile is statistically normal within a population of orbitofrontal neurons, we can treat be068 just like any other orbitofrontal neuron. A lack of 100 percent predictability is fine for the explanatory purposes of the vast majority of neuroscientific research studies.

But this attitude towards be068 or other nonhumans won't yield the metaphysico-semantic partition that the Technical-Behaviorist needs. A classical economist is free to adopt a Technical-Behaviorist interpretation of psychological ascriptions to humans if so desired. Take D-503 (the human protagonist in Zamyatin's dystopian novel *We*) as a classical economic agent. The fact that D-503's behavior is rational and predictable (in the sense of being more or less probable) does not deprive him of free will, second-order desires, emotions, neuroses, and so on. It's just that these features don't play a role in classical economic models. Of course, unlike be068, D-503 is *expected* to be capable of variability in behavior. But this expectation reflects our customary tolerances for variability set by our implicit or explicit explanatory purposes in relation to be068 or D-503. Classical economic models won't capture D-503's behavior perfectly, but neither will current models of neuronal behavior capture everything be068 does. We reveal our assumption of D-503's uniqueness by stressing his variability (despite the regularities of his behavior captured by the model) and reveal our assumption of be068's lack thereof by stressing its conformity (despite the exceptions to model predictions that it occasionally exhibits).

By varying our tolerances for variability, we can indeed develop and use new psychological polysemes, using one to refer to a pattern of behavior in some explanatory contexts and reserving the original term for cognitive states in other contexts. (There can be more than two polysemes, but the Technical-Behaviorist is only positing two.) But such polysemes can be developed and used for humans and nonhumans alike. Literalism is compatible with the development of psychological polysemes that reflect new biological knowledge. What she denies is that psychological ascriptions will be divided such that the behavior-referring polysemes are used for nonhumans and the original notions for humans. It's this matched partitioning of ascriptions into human/real and nonhuman/Intentional Stance that the Technical-Behaviorist must motivate and defend.

The Technical-Behaviorist might insist that human behavior is normatively guided. We are rational in the sense that we do things for reasons of which we are aware and to which we are responsive and held responsible for, whereas neurons are not aware of or responsive to reasons. Only normatively guided behavior requires explanation in terms of genuine psychological states. The Intentional Stance suffices for non-normatively guided entities.

We have also seen this move before (ISQ #8; see Chapter 4). But now the Technical-Behaviorist is rejecting the Intentional Stance for humans for reasons that are independent of her desire to distinguish human ascriptions from nonhuman ones. She has abandoned the practical rationality of the original Intentional Stance, which applied to human and nonhuman behavior, for an inflated sense of rationality as norm-guided responsiveness to reasons. Genuine psychological states now presuppose normative rationality, not mere practical rationality.[5]

The obvious problem is that if this were correct then in many psychological explanations of human behavior we would not be ascribing

---

[5] This begins to tread on the issue of the extent to which Dennett's personal/subpersonal distinction maps onto an intentional/nonintentional explanation distinction (e.g. Hornsby 2000). Neurons are parts of humans, so if human body = person body, then neurons are certainly subpersonal in the mereological sense. It doesn't follow that neurons can't be properly ascribed preferences unless "subpersonal" is not just a mereological notion (which it isn't, as I understand Dennett's view).

genuine cognitive states at all. Humans at some times, if not many times, would fall on the wrong side of her referential partition. In general, establishing her desired referential distinction by beefing up the requirements for genuine states will leave much of human behavior and cognition in the lurch. In addition, there's no guarantee that our best explanation of normatively guided behavior will confirm the Technical-Behaviorist's division of ascriptions of stance-y and non-stance-y predicates to humans and nonhumans, respectively. Suppose our best explanation of normatively guided behavior invokes psychological resources that can be Literally ascribed to nonhumans and humans, distinguishing cases by adding appropriate (and to some extent species-specific) evolutionary and situational constraints and goals. This doesn't need to be naturalistic, but naturalists are much more likely to develop this kind of explanation. In this case, we still would not be partitioning our uses of psychological predicates between humans and nonhumans in her desired way.

In sum, I do not find a non-question-begging motivation for the Technical-Behaviorist's referential distinction nor a non-question-begging principle for drawing it. The view seems to depend ultimately on the fact that we're human and we have a history of downgrading nonhumans. True enough, but hardly a reason to adopt her referential distinction as the best way to interpret the new uses. In Chapter 9 I'll consider whether this is reason enough for a pragmatic rejection of Literalism, setting aside whether it is the best semantic view.

7.4 The Second Variant: Exsanguinated Properties

The Technical view holds that psychological predicates are used literally for nonhumans, but the terms are distant polysemes, with distinct reference from when they are used for humans. The Exsanguinated Property variant claims that psychological predications to nonhumans pick out capacities or properties that are in some sense lesser or deflated versions of the human capacities or properties. The terms express attenuated concepts that refer to exsanguinated properties of nonhumans, not non-attenuated concepts that refer to full-blooded properties of humans. To use Dennett's phrase (Bennett et al. 2007: 88), the predicates ascribe "hemi-semi-demi-proto-quasi-pseudo" cognitive properties to the nonhumans.[6] The role of the human standard, semantically and metaphysically, is probably most explicit in this alternative to Literalism. The human-associated concepts and capacities are the "home" cases, the others "derivative" (Dennett 2010: 59).

The Exsanguinated Property variant may be elaborated by drawing on the idea, constitutive of homuncular functionalism, that psychological properties at the personal or organism level will be explained in terms of ever stupider properties at the subpersonal level or levels (Attneave 1961; Fodor 1968b; Cummins 1975, 1983; Dennett 1978b; Lycan 1981, 1987, 1988). In Dennett's terms:

We don't attribute *fully-fledged* belief (or decision or desire—or pain, heaven knows) to the brain parts—that would be a fallacy. No, we attribute an attenuated sort of belief and desire to these parts, belief and desire stripped of many of their everyday connotations (about responsibility and comprehension, for instance)... [T]he idea is that, when we engineer a complex system (or reverse engineer a biological system like a person or a person's brain), we can make progress by breaking down the whole wonderful person into subpersons of sorts [of] agentlike systems that have *part* of the prowess of the person, and then these homunculi can be broken down further into still simpler, less personlike agents, and so forth...
(Dennett, in Bennett et al. 2007: 87–8)

On this view, sentient creatures are a kind of corporate entity with a nested hierarchy of functions and subfunctions "that cooperatively go about the business of interpreting the stimuli that impinge on the corporate organism and of producing appropriate behavioral responses" (Lycan 1988: 5; Dennett 1978b: 123–4 describes the view in terms of nested flow charts). Explanations of the capacities of the whole posit "agentlike systems that have part of the prowess of a person" until one gets to "simpler, less personlike agents" that are "so stupid they can be replaced by a machine" (Dennett, in Bennett et al. 2007: 88). At least for Dennett, the process of attenuation involves stripping away "everyday connotations" from the concepts—as noted, in the case of belief and desire, connotations about responsibility and comprehension. This results in terms that refer to sort-of beliefs or desires, not full-fledged beliefs or desires.

---

[6] Should they still be called "cognitive"? Maybe not. But this won't really matter. The key claim is that they are exsanguinated versions of cognitive properties or capacities, not just different properties or capacities (as in the Technical-Behaviorist variant).

In terms of this familiar framework, the Exsanguinated Property view posits a metaphysical hierarchy of properties or capacities that are full-blooded or increasingly exsanguinated, and maps this deflationary metaphysical hierarchy onto a mereological hierarchy of whole–part relationships between entities. The corresponding semantic view implies that there may be many, increasingly distant, polysemes as the properties that hemi-demi-etc. psychological predicates refer to are increasingly exsanguinated. This raises the problem of determining when one has a new polyseme referring to a new exsanguinated property. How many stages of exsanguination are there, and how many of these have semantic consequences? Since we often use psychological terms for humans in ways that do not seem to require the full suite of everyday connotations, it would be helpful to know which connotations, when stripped away, leave behind a lesser capacity. Since the view fails for other reasons, I won't press this issue, even though it is not trivial.

For comparison, on the Literalist picture the terms refer to the same full-blooded properties wherever the properties may be found. There can still be a useful personal/subpersonal distinction, but it would be a matter of differences in how the capacities are deployed given the relevant affordances, not the capacities themselves. It would be a distinction without metaphysical implications for psychological properties or semantic implications for the predicates that denote them.

7.5 The Literalist Responds to the Second Variant

The Exsanguinated Property variant, understood as a version of homuncular functionalism, suffers from an obvious defect in that it doesn't apply to many relevant cases. Since it only applies to parts of wholes to which full-blooded ascriptions are made, it is compatible with full-blooded cognitive properties being ascribed to plants, fruit flies, and the like. So this variant would not block many relevant ascriptions in biology from being interpreted Literally. Moreover, in the case of communal organisms, such as bacteria, it would block full-blooded ascriptions to individual bacteria only if one is also ontologically committed to colonies as wholes, but not otherwise. In short, given the wide spectrum of cases, the view appears ad hoc.

The deeper problems are that the Exsanguinated Property view intro-duces, rather than dispels, mystery, and does so for no good reason. I'll consider these in reverse order. Note that the problems do not change fundamentally without a homuncular functionalist gloss, and instead just involve ascriptions to nonhuman wholes.

First, the problem exsanguinated cognitive properties were originally introduced to solve—the homuncular fallacy—is not a fallacy. I don't mean: it *is* a fallacy but it's being avoided. I mean: it's not a fallacy so there's nothing to avoid. I discuss this at length in Chapter 8, but the basic idea is that the same capacities are regularly ascribed at different levels in perfectly acceptable decompositional explanations throughout science, and there's no plausible way to carve out an excep-tion for psychological capacities. As a result, there is no need to posit "hemi-demi-semi-proto-quasi-pseudo" properties. This leaves the Exsan-guinated Property variant without its main motivation. Any other motiv-ation cannot merely point to the distinction between humans and nonhumans.

Second, what are these exsanguinated properties anyway? The meta-physics of exsanguination should provide more insight than just saying that these properties or capacities are not quite what humans have. The metaphor of ever "stupider" homunculi has not, to my knowledge, been cashed out in any detail (as Ramsey 2016: 4 fn. 1 also notes for specific cases). The analogy Dennett uses to illustrate the view in Bennett et al. (2007) is unpersuasive. He suggests that the attenuated concepts are akin to a child's sort of believing that her father is a doctor without fully understanding what "doctor" really means (Bennett et al. 2007: 87). But, as Stich (1983) argued, many adults can have beliefs about things they don't fully understand; the implied conditions for concept possession would deprive many of us of many of our concepts. Moreover, one's capacity to believe is not attenuated even if one's grasp of a particular concept is attenuated. Confused thoughts are still full-blooded thoughts. Concepts, attenuated or no, are not in dispute. Capacities are. If the issue is one of fewer options for exercising a capacity, that is compatible with a less extravagant metaphysics in which full-blooded capacities are exercised in context-dependent ways. A quick decision regarding the direction of coherent motion of dots on a computer monitor is not a hemi-demi-semi-proto-quasi-pseudo decision just because it is not as fraught as a decision whether to commit suicide. A nursing infant can only prefer one of two breasts—that's just the way it is.

If Hubel and Wiesel are picking out an exsanguinated *capacity* when they use "prefers" for neurons, one needs to show that what neurons don't or can't do is essential to full-fledged, real, or genuine preferring. That's the metaphysical issue. In semantic terms, one needs to show that what has been shaved off the meaning of "prefers" is essential to its meaning, akin to how IS MADE OF FLOUR might be considered essential to "pancake". It is not enough to say that IS RESPONSIBLE or COMPREHENDS are normally contextually associated with "prefers". This may or may not be true, and is compatible with Literalism. It must show that these features are part and parcel of its lexical or encoded or context-independent meaning. We need a non-question-begging reason to think the features indicated by "everyday connotations"—which are no doubt anthropocentric—are metaphysically determinative of the properties, such that when stripped away they leave behind only anemic versions of the real thing.

The Exsanguinated Property proponent might respond that the cases of semantic revision discussed at the end of Chapter 3 are actual examples of the attenuated concepts she has in mind. For example, "surprisal" is distinguished from "surprise" by the fact that SURPRISAL leaves out some aspect of emotional response to the unexpected. Perhaps SURPRISAL has some semantic overlap with SURPRISE, but they are referentially distinct. The former picks out an exsanguinated version of the property picked out by the latter, and parts of wholes (or nonhuman wholes) that are ascribed "surprise" actually pick out SURPRISAL. Similarly, *mutatis mutandis*, for PREFERS when ascribed to humans and neurons, even though in this case we lack a newly coined polyseme expressing the exsanguinated concept. Moreover, relevance-theoretic semantics (Chapter 6) offers a natural way to articulate a semantics of attenuated concepts. The Exsanguinated Property variant's continuum would not run from literal to metaphorical use, but instead would be a continuum of literal uses ordered by the full-blooded to exsanguinated nature of the capacities to which the terms refer. All this helps unpack the metaphor of "stupider" homunculi and provide real examples of it.

But RT does not say which aspects of a capacity that humans have are essential to the capacity, and intuitions differ (e.g. the LOYALTY example; see Chapter 6). The issue here, however, is not that we can't distinguish literal from metaphorical use without begging this question. What is missing now is a reason to classify surprisal (or SUR-PRISAL) and its even more anemic cousins as exsanguinated relative to surprise (or SURPRISE) that does not take the naïve human standard for full-bloodedness as given. We also need a reason to map this series of increasingly exsanguinated properties to a series of mereologically related entities or a series of increasingly less human entities (a kind of Great Chain of Being). More precisely, the semantic continuum of attenuated terms (or concepts) must be mapped to a metaphysical continuum of exsanguinated properties *and* to a mereological (or increasingly less human) continuum of entities. All three continua must line up in the right way, with humans at the full-blooded end of the property continuum and the non-attenuated end of the semantic (or conceptual) continuum. This is as clear a representation of the anthropocentric semantics of the Manifest Image as one might hope to find.

Of course scientifically driven referential revision can result in distinct kinds, neither of which is the metaphysical standard for the other (e.g. attention and awareness). The Technical-Behaviorist view falls in this category. The human-centered concepts and capacities are sequestered in their original home without enlisting them to clarify the nature of what nonhumans have. Surprisal is just a behavioral pattern.

Scientifically driven referential revision can also result in a newly precise standard for a kind. Literalism falls in this category. Human-centered concepts and capacities are adjusted in the light of new evidence to make them not human-centered. Science affects us too. Surprisal is surprise properly understood. The unexpected may or may not generate the emotional response we often associate with it based on some of our own experiences.

In contrast to these responses to scientific discovery, the Exsanguinated Property view posits a series of sort-of distinct kinds that remain forever anchored to the human case as traditionally conceived. Human-centered concepts and capacities are adjusted to fit nonhuman cases, but without abandoning the human standard for the nonhuman cases. Apparently many find this view intuitively compelling. At least it assuages worries about naturalizing the mind (but see Chapter 8). But without non-metaphorical illumination of the metaphysical exsanguination and seman-tic attenuation processes, plus reason to think these processes will line up with mereological (or Great-Chain-of-Being) relations, we cannot know whether it can avoid begging the critical questions raised in this book.

7.6 Concluding Remarks

This chapter concludes my discussion of the semantic alternatives to Literalism. On either variant, the Technical view fails to draw her required referential distinction between predications to humans and nonhumans in a non-ad hoc, non-question-begging way. The Technical-Behaviorist variant seems initially plausible, but it turns out to be difficult to motivate and defend. The Exsanguinated Property variant doesn't even get that far. It introduces metaphysical mystery and it assumes the anthropocentric standard that is in dispute. The human case, with all everyday bells and whistles intact, is the standard for full-bloodedness and capacities of nonhumans are exsanguinated relative to this human standard.

Between the views presented in Chapters 5 and 6 and those presented here, I have discussed what I believe to be the non-Literalist gamut. None of these alternatives even begins to approach Literalism in providing a non-ad hoc explanation of the range of uses that need to be explained, a small sample of which was exhibited in Chapters 2 and 3. I cannot prove that no genuinely distinct other alternatives are possible, let alone plausible, but I leave the challenge of actually coming up with one to the anti-Literalist. Other positions would almost certainly be variations on one or another of these basic themes, or would face very similar problems when articulated.

In contrast, the benefits of Literalism are many. (i) It shows biologists as responding rationally to their accumulating knowledge: they use psychological predicates to report and theorize about their discoveries just as they do using predicates appropriate for describing other discoveries. These predicates are as apt for extension and revision in the light of science as any other type. (ii) It saves the appearances of literal use. The predicates are used in just the way other terms are used in contexts when literal interpretation is the default. (iii) It promotes empirical inquiry and clarity. Findings of similarity across domains encourage further research and consistency of interpretation helps avoid confusion. (iv) It does not introduce mystery into our explanations of the mind. We may not know a lot about the mind, but it does not help matters to introduce mysterious new properties or insist that intuition is our best and most reliable source of evidence. (v) It meshes with the independently known extensional flexibility of verbs and their literal ascriptions across object domains and at different levels within object part–whole hierarchies. This point will be discussed further in Chapter 8. (vi) Finally, it helps us appreciate that we occupy front-row seats for observing a very important case of adjusting language to reality. What is happening to psychology in the light of biological research is exciting. Literalism embraces these advances. What's not to like?

In Chapters 8 and 9 I consider two reasons. These involve the apparent implications of Literalism for mechanistic explanation of the mind and for the traditional assumption that possession of the cognitive capacities now in play are what make us moral agents and moral patients. Does Literalism entail that no mechanistic explanation of mind is possible? Does it entail that we do not deserve the special treatment worthy of our status as persons? Each of these implications, if true, would provide a pragmatic reason to seek an alternative to Literalism or shore up one already presented. In Chapters 8 and 9, I will argue that Literalism has no such abysmal consequences either for naturalism about the mind or the ways we treat humans. The sky will not fall.

8

Literalism and Mechanistic Explanation

8.1 General Remarks

The idea that psychological predicates are literally true of entities in many more domains than lay intuition accepts may appear incompatible with explaining the mind in terms of mechanisms (or, more generally, in naturalistic terms). If Literalism is correct, some of the entities that could have such capacities (in principle) are neurons or brain structures. Neurons and brain structures are *bona fide* explanantia in actual mechanistic explanations of brain-based minds. But a longstanding criterion for an adequate explanation of mind, familiar from homuncular functionalism, holds that the ascription of psychological capacities of wholes to their parts is non-explanatory. An explanation of intelligence that posits intelligent components is no explanation at all. Literalism would be severely undermined if it entailed committing the so-called homuncular fallacy. The objection seems to show that Literalism and mechanistic explanation of brain-based minds don't mix.

As a scope restriction, the objection would not do much to undermine Literalism, which does not and need not claim that psychological properties are in fact possessed by neurons or brain structures in order to be plausible. Neurons don't necessarily have such capacities, and current ascriptions could be false. I've repeatedly used the ascription of preferring to neurons for argument's sake, but I also noted that this particular case might not be true. Nevertheless, I agree it is important to show that Literalism is compatible in principle with mechanistic explanation of brain-based minds.[1] Even further, I will show how it promotes a better

[1] I don't claim that mechanistic explanation is the only form of explanation or the only form of a naturalistic explanation of mind. The issue of how other explanatory practices,

conceptual framework for naturalism than the framework from which the homuncular fallacy emerges.

In what follows, I will elaborate the objection in the context of homuncular functionalism. I then present contemporary mechanistic explanation and its neutrality regarding proper levels for ascribing activities within part–whole object hierarchies. I will argue that the homuncular fallacy is not a fallacy in general. Many activities are not specific to particular levels of object composition, and the predicates that pick them out are not proprietary to the sciences that focus on those levels.[2] Moreover, at least *prima facie*, psychological capacities are not treated as exceptions in scientific practice. I consider and reject various ways to justify a psychological exception, and provide an alternative interpretation of the homuncular functionalist's demand that an adequate naturalistic explanation of the mind must "discharge" psychological concepts.

8.2 Homuncular Functionalism and Mechanistic Explanation

Prior to recent work in mechanistic explanation, the idea of a decompositional explanation of mind was articulated in philosophy of mind in the form of homuncular functionalism (Fodor 1968a; Cummins 1975, 1983; Dennett 1975a/1978b; Lycan 1981, 1987, 1991; in psychology, Attneave 1961). On this view, one explains a mental capacity of a whole in terms of the organized operations of its parts, with two essential and characteristic caveats: the subcapacities ascribed to the parts must be

such as mathematical, computational, or dynamical systems modeling, are related to mechanistic explanation is discussed in the text.

[2] In this chapter I'll often use the term "activity", as well as "capacity" or "property", because this term is used frequently in the literature on mechanistic explanation for the broad ontological class of what Simons (1987, 2000) labels "occurrents" (processes, events) in contrast with "continuants" (objects). Following Machamer et al. (2000: 30), capacities, dispositions, tendencies, propensities, powers, or endeavors are derivative from activities in that individuating and identifying the former presupposes individuating and identifying activities (e.g. a disposition to φ depends on what it is to φ). My discussion does not depend on their commitment to quantifying over activities. See Steward (1997) for discussion of the ontology of mind in terms of metaphysical categories of events, processes, and states (which she does not consider a unified category), and Galton and Mizoguchi (2009) for an ontology that puts neither objects nor processes first.

distinct from and simpler than the capacity of the whole that they explain. Expanding on Attneave (1961), Lycan states the homuncular functionalist orthodoxy as follows:

> It was Attneave's insight that homunculi can after all be useful posits, so long as their appointed functions do not simply parrot the intelligent capacities being explained. For a subject's intelligent performance can be explained as being the joint product of several constituent performances, individually less demanding, by subagencies of the subject acting in concert. We account for the subject's intelligent activity, not by idly positing a single homunculus within that subject whose job it simply is to perform that activity, but by reference to a collaborative team of homunculi, whose members are individually more specialized and less talented. The team members' own functions are specified first, and then the explanation details the ways in which the members cooperate in producing the more intelligent corporate explanandum activity.   (Lycan 1991: 260)

The call for subfunctions that are "individually less demanding" (see also Lycan 1987: 40) is echoed by Cummins (1983: 28–38 passim) and Dennett (1975a/1978b: 80), who describe them variously as "less problematic", "simpler", "elementary", "primitive", or "less clever".

These two restrictions on permissible explanantia of psychological capacities in a decompositional framework are motivated by the worry that ascribing the capacity of a whole to a part is non-explanatory.[3] Lycan, cited above, claims that without these restrictions the explanation would be idle. Bechtel (2009: 561) concurs that "assuming a homunculus with the same capacities as the agent in which it is posited to reside

---

[3] This requirement is sometimes seen as a response to a regress problem raised by Ryle (1949) against the "intellectualist legend" (Lycan 1988: 40; Ramsey 2007, 190) that:

> whenever an agent does anything intelligently, his act is preceded and steered by another internal act of considering a regulative proposition appropriate to his practical problem.... [T]he absurd assumption made by the intellectualist legend is this, that a performance of any sort inherits all its title to intelligence from some anterior internal operation of planning what to do.   (Ryle 1949: 29–32)

That is, an act can only be intelligent if the prior planning process is itself intelligent, which itself must be preceded by another intelligent operation if the inheritance relation is to hold. This infinite regress "reduces to absurdity the theory that for an operation to be intelligent it must be steered by a prior intellectual operation". But Ryle's regress is not essentially about homunculi (i.e. wholes and parts) at all. The legend holds that (1) an *inner* operation or act precedes an *outer* one and (2) the outer one counts as intelligent by inheritance from the intelligence of the inner one. Both operations (the public behavior and the private planning capacity) may be ascribed to the whole system.

clearly produces no explanatory gain". Dennett (1975a/1978b: 72–73, 83) describes the proper sort of explanatory procedure as follows:

[T]he account of intelligence required of psychology must not of course be question-begging. It must not explain intelligence in terms of intelligence, for instance... by putting clever homunculi at the control panels of the nervous system.... A non-question-begging psychology will be a psychology that makes no ultimate appeals to unexplained intelligence, and that condition can be reformulated as the condition that whatever functional parts a psychology breaks its subjects into, the smallest, or most fundamental, or least sophisticated parts must not be supposed to perform tasks or follow procedures requiring intelligence.

This procedure guarantees that eventually the homunculi are "discharged" in the sense that no questions about intelligence are begged. The discharging process can involve as many intermediate homunculi as needed, so long as none of them quite reaches the "full-blooded intentionality of the human mind" (Lycan 1987: 58). The exsanguinated properties discussed in Chapter 7 are the homunculi required for decompositional explanation of mind so conceived.

The explanatory restrictions of homuncular functionalism are echoed in a potent epistemological history of rejection of vitalism and the original homunculi in the face of mechanistic explanations of nature. As first proposed, homunculi were preformed little men that were posited to explain, within the Aristotelian tradition, how the fully formed adult could emerge from an embryo (Maienschein 2017). Heaping ridicule on homunculi was part of a general rejection of this explanatory tradition in favor of a new mechanistic view in which life and growth could be explained in terms of direct contact motion. Molière's reference in *Le Malade Imaginaire* to the dormitive virtue of a sleeping medicine is an oft-cited literary example of this anti-Aristotelianism. Leibniz too lampooned people who "saved the appearances by explicitly fabricating suitable occult qualities or faculties, which were thought to be like little demons or spirits able to do what was required of them without any fuss, just as if pocket watches told time by some faculty of clockness without the need of wheels, or mills crushed grain by a fractive faculty without the need of anything like millstones" (Preface to the *New Essays* (1703–5), in Ariew and Garber 1989: 306).[4] Fodor (1968a: 627) recapitulates the

---

[4] See also Dennett (1978b: 56).

historical ridiculing of homunculi for contemporary readers with a farcical computational account of how to tie one's shoes in terms of a little man inside who follows a rulebook for tying one's shoes.

For some (e.g. De la Mettrie), explanation by means of push–pull mechanisms also included the mind. Others, such as Leibniz, held that no mechanistic explanation of mind (or, at least, of perception) is possible:

17. Moreover, we must confess that the *perception*, and what depends on it, *is inexplicable in terms of mechanical reasons*, that is, through shapes and motions. If we imagine that there is a machine whose structure makes it think, sense, and have perceptions, we could conceive it enlarged, keeping the same proportions, so that we could enter into it, as one enters into a mill. Assuming that, when inspecting its interior, we will only find parts that push one another, and we will never find anything to explain a perception. And so, we should seek perception in the simple substance and not in the composite or the machine.

> (Leibniz, *The Principles of Philosophy, or the Monadology* (1714),
> in Ariew and Garber 1989: 215)

Homuncular functionalism sides with De la Mettrie that the mind can be explained in terms of the organized operations of parts. But it reconceives homunculi as proto-cognitive operations rather than preformed little people, and adds that they form a nested series of ever-simpler operations that ends with fully non-cognitive operations. As Fodor (1968: 629) puts it, a homunculus is a "representative *pro tem*" for a system of instructions "that makes no reference to unanalyzed psychological processes". Ever-stupider homunculi enable us to satisfy the conditions for explanatory adequacy.

Like homunculi, mechanistic explanation has also been updated from its original form. In general contemporary terms, a mechanistic explanation is an explanation of a capacity of an entity in terms of an entity's constituent entities and activities (Machamer et al. 2000; Craver 2007).[5] Importantly,

---

[5] Articulations of contemporary mechanism also include Bechtel and Richardson (1993), Glennan (2002), Bechtel and Abrahamsen (2005), and Bechtel (2008). For some, the mind is not subject to mechanistic explanation for normative reasons (Kaplan and Craver 2011: 603 fn.1, citing McDowell 1994), but the question of normativity of mind is independent of mechanistic explanation. Mentality could be normative wherever it is properly ascribed; the key is to not build "human" into "normative" (see Chapter 4). Extended mind theorists can explain the (human) mind mechanistically, just not depending solely on brain function. My discussion is neutral on the issues of whether all explanation is mechanistic (Chemero and Silberstein 2008; Kaplan and Craver 2011; Levy and Bechtel 2013; Chirimuuta 2014), as well as how to distinguish mechanistic explanations, as opposed to, say, detail-challenged

contemporary mechanists do not restrict mechanistic explanation to "exclusively mechanical (push-pull) systems" (Machamer et al. 2000: 2). They emphasize the characteristic explanatory method rather than characteristic types of motions or operations. Kaplan and Craver (2011: 605) summarize this difference: "[T]o insist on mechanistic explanations is not to insist on explanation in terms of simple machines governed by strict deterministic laws or in terms of physical contact, energy conservation, or any other fundamental or otherwise privileged set of activities." Just so, the main alternatives to the new mechanism are deductive-nomological explanation, dynamical systems theory, and computational modeling (although these are not necessarily opposing or non-overlapping views). The lack of restriction to push–pull mechanisms is justified by the fact that the new mechanistic model is crafted to capture actual mechanistic explanatory practices in science. As Craver (2007: 4) puts it: "One can restrict the class of machines to heroic simple machines (levers, pulleys, and screws), or to extended things colliding (as in Cartesian mechanism) or to things that attract and repel one another (as du Bois Reymond held). Each of these restrictions makes the concept of a mechanism too narrow to accommodate the diverse kinds of mechanism in contemporary neuroscience." Craver's moral extends across contemporary sciences engaging in mechanistic explanation as well.

These same explanatory practices that make restriction to push–pull mechanisms obsolete also show that many accepted mechanistic explanations involve the same capacities being ascribed to wholes and their functional parts. A piece of machinery lifts boxes because its engine has a camshaft that lifts valve covers. If a piano string is vibrating (exhibiting simple harmonic motion), this will be in part because the molecules in the string also vibrate. Exchanges of goods and services between countries are composed of such exchanges between private and public agencies or companies and, ultimately, individual people. The rotational motion of the planets explains how our solar system maintains its stable dynamic equilibrium, the rotational motion of the solar system explains how it maintains its stable dynamic equilibrium relative to other such

---

mechanism sketches, and in general how much detail is required for a *good* explanation or any *explanation* at all (e.g. Levy and Bechtel 2013). Even if all explanations or models involving psychological predicates are just mechanism sketches (Piccinini and Craver 2011), nothing follows about where the predicates are properly ascribed.

systems in our galaxy, and the rotational motion of galaxy clusters keeps them from collapsing into each other. A monkey learns to swing through the trees by grabbing and releasing vines, and presynaptic neurons in his hippocampus release glutamate in the process of long-term potentiation, theorized to be a mechanism of learning. If my opening my mouth is in part explained by the valves in my heart opening and the ion channels in my cardiac cells opening, I have explained my mouth's opening and I have referred to opening in the explanans (twice!). As the nursery rhyme goes:

> Big fleas have little fleas
> Upon their backs to bite 'em
> And little fleas have lesser fleas
> And so ad infinitum.[6]

In short, it is common for the same kinds of activities to occur at different spatiotemporal scales and to comprise in part the activities they help mechanistically explain (although the point is neutral on the issue of ontic explanation). The fact that different entities do the same things (perhaps in different ways) does not entail either a difference in activity kind or that a performance at one level is real or full-blooded while the others are not. It also does not undermine the quality of the mechanistic explanations in which such repetition occurs.

There is also explicit disregard of the homuncular functionalist restrictions within areas of biology and psychology relevant to the present debate. As already noted (Chapter 3), Sutton and Barto (1981) theorized about adaptive elements that would comprise adaptive systems, intending their model of reinforcement learning to apply to both (see also Barto 1995: 4–6 on the actor-critic model). Ben Jacob et al. (2006: 517–18) suggest that semantics may be generated at multiple levels of an organism's hierarchy of function. This is an armchair conjecture, but they show no concern that their proposal might lead to epistemic disaster.

---

[6] See <https://en.wikipedia.org/wiki/The_Siphonaptera> for such rhymes. Thanks to John Norton for the reference (involving whorls), although presumably fleas will never be included among the fundamental entities of the universe. Also, none of the examples cited in the text is a complete explanation. Due to scale differences and discontinuities, actual explanations will not be this direct, and we may never bother to articulate them fully. This just means that the core decompositional commitment of mechanistic explanation, as illustrated by Craver's (2007: 7) flat-topped pyramid diagram, is too simple to show how all-the-way-down mechanistic explanations would really go.

Trewavas, one of the leading figures in plant intelligence, considers the relevant spatiotemporal scales of plant behaviors to run from individual cells to entire plants, which "runs counter to the idea that 'behavior' is something reserved for whole organisms" (Zink and He 2015: 724, reviewing Trewavas 2014). Wimsatt (2006: 461) remarks rhetorically: "Memory—a property of molecules, neural circuits, tracts, hemispheres, brains-in-vats, embodied socialized enculturated beings, or institutions?" Wimsatt scoffs at levels in nature, but even if one likes levels it is clear that the practice of ascribing the same capacities across them explicitly includes or does not explicitly exclude psychology.[7]

This rejection of the homuncular functionalist's adequacy conditions is not entirely new. Margolis (1980) argued that there is no a priori reason to restrict explaining capacities to stupider ones. Freudian psychology posited a complex unconscious which explained simple or complex patterns of behavior. So non-identity suffices to avoid the alleged epistemic problem. Sober (1982: 420) adds that there is no general scientific reason even to require non-identity: "What, then, is wrong with smart homunculi? Why must homunculi be stupider than we are, if 'Chinese box explanations', as we might call them, are permitted in other domains?" On his view, the apparent "emptiness" of homuncular explanations does not stem from ascriptions of the same operations to parts and wholes, but from a shift from tokens to types: e.g. even if planets rotate in part because their atoms rotate, we are no closer to an explanation of rotating ("what unites planets and atoms") (1982: 421). As I'll discuss further below, this is not just any token–type shift. It is a shift from explaining objects (planets and atoms) to explaining activities (rotating), which motivates research into activity kinds—e.g. what rotating is. Presumably our best explanations of activity kinds will also show why ascribing the same activities at multiple levels in a part–whole object hierarchy does not yield an "empty" or idle explanation.

The homuncular functionalist restrictions on allowable explanantia in a decompositional explanation of mind are absent from contemporary models of mechanistic explanation and are ignored in contemporary mechanistic explanations in other domains. There is no general explanatory

---

[7] Wimsatt (1976: 253) illustrates a "bio-psychological thicket" in which levels-talk is idealized (or maybe counterproductive or just plain silly). Churchland and Sejnowski (1988: 742) provide a classic levels picture within neuroscience.

fallacy for Literalism to run afoul of. The alleged fallacy doesn't seem to bother cognitive scientists either. Literalism provides a straightforward explanation of this lack of concern about psychological ascriptions to parts. So why think decompositional explanations in psychology are epistemically exceptional? Why do we need homunculi?

Perhaps the metaphor of "stupider" homunculi can be interpreted as a way of gesturing at the idea that we don't want to ascribe all aspects of human manifestations of cognitive capacities to the parts, *without* implying that by leaving these aspects out what is ascribed to the parts is anything less than full-blooded. (This would not be Dennett's view, however: Bennett et al. 2007: 87.) On this interpretation, the "same capacity" would be one we have, but that our parts can also have in just as full-blooded a manner. Just so, the Literalist does not claim that neurons prefer what humans do or exhibit their preferences in all the ways humans exhibit theirs. They both full-bloodedly prefer what they do in the ways they do. I have no objection to homuncular functionalism interpreted as closet Literalism, but if reference to ever-stupider homunculi implies that the whole-human case remains the standard for all real cognitive capacities, it cannot be Literalist.

8.3 Seeking a Psychological Exception

Nevertheless, it would be odd for the homuncular fallacy to have had such a grip in psychological and philosophical circles if there were nothing to it. We need to consider whether the psychological exception can be justified.

Perhaps the problem arises in psychology due to incompleteness. In one sense, a mechanistic explanation can be incomplete because we still need to explain a capacity at one level in terms of capacities at lower levels (without implying that "lower-level" capacities are level-specific). But while an explanation that is incomplete in this sense may not be wholly satisfactory, it is not thereby idle, circular, or question-begging.[8] We knew that ion channels in neural and cardiac cell membranes opened

---

[8] How this point is expressed may depend on where one draws the line between an explanation and something that for some reason does not count as an explanation. For the sake of exposition, I assume the class of explanations includes things that are not complete, by some justified criterion of completeness such that incomplete explanations are possible.

long before Reuveny (2013) discovered exactly how. Even as we sought a more complete explanation of neural signaling or heart function, what had been said was not epistemically faulty. Explanations that ascribe psychological operations to parts are no doubt incomplete in this sense, but it is not clear why this incompleteness would be epistemically damaging.

The homuncular functionalist might respond that such ascriptions would be as much of an advance towards the goal of explaining the mind as ascribing a dormitive virtue to a sleeping pill (or ascribing a little internal man who follows a rulebook for tying shoes to explain how the big man ties his shoes). Wanting a more complete explanation doesn't excuse us from the requirement to avoid idle, non-illuminating moves along the way.

But dormitive virtues lost their explanatory power because virtues no longer had explanatory force within the original mechanistic explanatory framework. They became idle posits. Rather than an argument for an exception, the objection can be turned against the homuncular functionalist. Modern-day homunculi can go the way of the dormitive virtues because psychological explanations in the contemporary context do not require homunculi.

Psychological explanation as conceived in homuncular functionalism is closely tied to early AI and what is now known as classical computationalism (e.g. Newell et al. 1958; Miller et al. 1960; Dennett 1975a/1978b). This dominance has eroded with the steady, if steadily disputed, rise of connectionist and dynamical systems approaches, including computational modeling, machine learning, and network science. If mechanistic explanation of brain-based minds is possible, it must relate psychology to neuroscience. In contemporary terms, however, this is a special case of the general problem of how mathematical models or other systems-level explanations are related to biological mechanisms. The use of the same models and other systems-level frameworks at multiple mereological levels is a pervasive, rigorously pursued feature of contemporary science, including cognitive science (e.g. Alon 2007; Bullmore and Sporns 2009, 2012; Baronchelli et al. 2013; see also Chapter 3). Above I gave numerous examples of ascriptions of the same activities to parts and wholes in the same mechanistic explanatory hierarchies. In at least some of these cases, sameness is a matter of using the same equations at different scales— for example, the same equations of angular momentum or harmonic oscillation at different scales and for different objects. Requiring everstupider homunculi in this explanatory context is like insisting that only planets full-bloodedly rotate even though the same equations of angular momentum apply to their atoms. There are plenty of problems to solve to integrate psychology and neuroscience, but the alleged idleness of using models at multiple scales isn't one of them.

The homuncular functionalist might reply that psychological predicates change their reference when cognitive models are used for parts, but this response would just bring us back to Chapter 7 (or point to some other non-Literalist interpretation). One would have to ignore the fact that there is no such automatic change in reference for non-psychological predicates in models ascribed at multiple scales. The claim of a psychological exception needs defense in whatever form it takes.

But this still leaves the question of whether the use of models at multiple scales is illuminating. To use Sober's example, what is missing from the Chinese box explanation of rotating planets and atoms is an illuminating account of what rotating is, at any level at which it occurs. Of course, repetition can indeed be illuminating. We frequently import prior understanding of activities to ensure the intelligibility of many mechanistic explanations. If proteins in ion channels in cell walls twist open, we understand what the proteins are doing because we understand what it is to twist open, albeit not by observing cellular proteins. To paraphrase Dennett (2010: 49–50), you could hardly speak of an ion channel opening without prior cases of opening to serve as exemplars. And if there is no such prior understanding, an activity ascription is not illuminating at any level at which it is ascribed. In addition, illumination through repetition is an important element in epistemic metaphor, which includes ascribing operations to parts that are *at least* analogous to what the wholes do and may be the same type. As noted in Chapter 6, such ascriptions are entirely compatible with Literalism.

Moreover, models do provide illumination. Sober (1982: 421–2) puts the point in terms of laws, although the point generalizes to mathematical models. The equations of angular momentum illuminate what rotating is, at any level. It doesn't matter whether what's rotating is an atom, a ballerina, or a planet, or whether our understanding of rotating began with seeing ourselves turn around in the same spot. We'll know when new things are rotating and when things we think are rotating really aren't. In short, we discharged rotating by finding the right laws. We didn't get rid of "rotating" (or ROTATING).

Similarly, if we are told an organism digests in part because parasites in it digest, "we now want to know what laws govern the way organisms obtain energy from their environments, and how those laws apply simultaneously to hosts and the parasites they house". In the course of defending the explanatory power of mathematical models, Cartwright (1983: 145) provides an example in terms of the harmonic oscillator model. This model "is used repeatedly in quantum mechanics, even when it is difficult to figure out exactly what is oscillating: the hydrogen atom is pictured as an oscillating electron; the electromagnetic field as a collection of quantized oscillators; the laser as a van der Pol oscillator; and so on. The same description again and again gives explanatory power." The model provides understanding of the behavior of the real systems modeled, whatever their level. Neither of these cases on its own involves explanation by decomposition, but as noted models and mechanisms are now used conjointly in explanations of behavior, without regard to the homuncular functionalist's explanatory constraints.

The homuncular functionalist might say that psychological capacities are special because, unlike rotating or oscillating, we don't understand psychological capacities to begin with. This reply conflicts with the fact that homunculi are intended to help with comprehension, even if "the aid to comprehension of anthropomorphising the elements" (Dennett 1975a/ 1978b: 81) is considered a temporary epistemic crutch. (The reference to "anthropomorphising" presupposes the anthropocentric standard in dispute.) Homunculi could never play their intended role in homuncular functionalist explanations if we did not have *some* understanding of the full-blooded properties to get a grip on the fractionally-blooded ones. It is also implausible to interpret homuncular functionalists as claiming we know *nothing* about psychological capacities from our own case. Not even the Literalist makes this claim. But if we do know *something* about them, psychological ascriptions to parts are as illuminating and legitimate as any other extension in science.

But the homuncular functionalist may insist that in the cases of rotating and the like we have physical concepts that explain other physical concepts. In psychology, we must at some point cross a conceptual boundary that need not be crossed in other cases. The Literalist will caution that the issue is not whether mental concepts denote capacities that are special, but whether the special capacities they denote are restricted to us (or wholes). But the homuncular functionalist will add that if we ascribe these capacities full-bloodedly to parts, even via models, such ascriptions are devastatingly idle because we want to understand them *in non-mental terms*. We do not want unanalyzed, spooky psychological processes in a naturalistic explanation of the mind. We need to *discharge*, not just mechanistically explain, these processes. Discharging is overcoming this conceptual distinction between the mental and the physical within the mechanistic explanatory context. None of the explanations above involve *discharging* because they don't need to cross this conceptual divide. They do not require *this kind of* illumination. A psychological explanation that discharges the mental will not just leave us with the same concepts we began with. Ever-stupider homunculi are indispensable for this purpose.

8.4 Discharging Discharging

But the homuncular functionalist's conception of discharging is not the only one possible. Dennett describes discharging as when there are no questions about intelligence being begged. But the homuncular functionalist cannot require that discharging entails explaining the mental in homuncular fashion. For the Literalist, ascribing a psychological capacity to a neuron is no more idle or question-begging than ascribing rotating to an atom. We need an understanding of discharging that does not beg the question against the Literalist.

We can agree that the mind is spooky in the sense that we don't understand it very well yet. I can also agree that discharging the mind means explaining it in a way that leaves it non-spooky. But the homuncular functionalist puts forward a very specific interpretation of what non-spookiness requires. On her view spookiness fades as psychological concepts and the capacities to which they refer are phased out by means of ever more attenuated concepts that refer to ever stupider homunculi (i.e. ever simpler capacities) (see Chapter 7). This view implies that the concepts and/or capacities are to blame for the spookiness. But why scapegoat them? To eliminate spookiness, what we really want is a way to get at the psychological from another perspective than the one we're starting from. We want a different way to identify and understand what the terms refer to or what the concepts are about.

This requirement can be articulated by saying we need to explain the mental in non-mental terms. Homuncular functionalism takes this requirement literally. It doesn't matter whether the concepts or the capacities come first. We can incrementally shave off informational components from the non-attenuated concepts and posit ever simpler capacities as the new referents, or we can incrementally shave off features from the full-blooded capacities and create hemi-demi-semi concepts to refer to them. Either way, mentality is conceptualized in terms of what humans have, and non-mentality results from gradually receding from this metaphysico-semantic standard. I find this mysterious and unjustified (see Chapter 7), but the current issue is whether this is the only way to discharge the mind. It isn't. We don't need to take this explanatory requirement literally as a matter of finding terms (or concepts) to replace the mental ones and positing a cascade of exsanguinated properties as their referents.

We now have, or can plausibly hope to find, mathematical models that can discharge the mental capacities without eliminating mental concepts. We don't need a series of increasingly attenuated pseudo-mental concepts and increasingly simpler capacities to do the job. Cognitive models can provide the desired non-anthropocentric perspective on the mental capacities precisely because their domain of application includes the nonhuman. Mental concepts are as apt as physical concepts for extension to any level because models and their construals do not belong to levels except contingently. For the Literalist, there is a special epistemic problem for explaining the mind: to distill from our homegrown understanding of the psychological which features are specific to us but contingent to possession of the capacity. Models, not homunculi, are indispensable for solving this problem.[9] As their name indicates, homunculi are infested with the human. They are about as helpful in this endeavor as the virtues.

The homuncular functionalist may reply that this way of discharging the psychological ignores the mechanistic picture. The core idea of decompositional explanation is not just that the activities of parts explain the activities of wholes, but that these lower-level activities are more basic or simpler (*pace* Margolis' 1980 objection). If the same psychological capacities are ascribed to parts, am I suggesting that no mechanistic explanation of mind is possible, that *psychological* capacities are basic

---

[9] Dennett (1975a/1978b) himself posits the nesting or repetition of generate-and-select procedures in any possible psychological explanation of adaptive behavior. If the Law of Effect were refashioned into a model, the Literalist would say that the standard construal of the model should be uniformly interpreted wherever it is successfully used. If it is a cognitive model, it doesn't stop being a cognitive model when it is used in a new domain.

or simple (or unanalyzable)? Am I siding with Leibniz rather than De la Mettrie?

I'm not sure this is an objection so much as a *cri du coeur* in the face of the possibility that received views of what naturalization requires are mistaken. The use of the same (or different) models at multiple spatial and temporal scales is integrated into the practice of explaining the behavior of wholes in terms of the behavior of their parts. As an objection, however, it may reflect a metaphysical confusion that stems from not adequately distinguishing objects from activities in mechanistic explanations. (A related issue regarding nominal and verbal metaphor arose in Chapter 6.)

Take the idea of basicness, and assume for argument's sake that levels-talk is coherent, metaphysical, and mereological. On the received view of levels, levels involve entities composed of other entities. The ultimate composing entities are the basic ones. Levels-talk so understood is inappropriate for elaborating the concept of a metaphysically basic activity. If planets, ballerinas, and classical-mechanical atoms rotate, there is a derivative, metonymic sense in which the activity-type rotating is basic: an activity is metonymically basic if and because basic *objects* are among the things that happen to be able to do it. Rotating is certainly not basic because we discovered rotating first in atoms, or because only atoms rotate. If psychological properties are basic in this sense, it would be if and because basic objects have them.[10] Literalism does not claim that psychological properties are basic in this derivative sense. But even if they were, it's certainly compatible with mechanistic explanation of mind to discover that the basic objects have psychological properties. The properties would then be basic metonymically, but not necessarily basic metaphysically.

The same moral holds for simplicity or analyzability. Assume for argument's sake that simplicity or analyzability involves decomposing a

---

[10] There are other senses of "basic" that will not matter here. We have a preferred level of activity ascription that is basic in the Roschian sense of a preferred conceptual level for categorization (Rosch 1973, 1975; Rosch et al. 1976; Figdor 2017). For example, in object categorization, "dog" is basic, rather than "mammal" or "Dalmatian". This sense of "basic" will not pick out all and only the metonymically basic: "eating" is basic in this sense. Also, in action theory some actions are "basic" because they are what we do with our bodies, as opposed to what we do *by* performing a basic action. In this sense of "basic", a ballerina's rotating is basic: she dances *by* rotating (in part).

complex into simples which cannot themselves be decomposed. The concept of simplicity also does not apply to activities the way it applies to objects. A mechanistic explanation of a ballerina's rotating is an explanation of how a human being with sufficient training rotates. This way of rotating requires arms and legs and takes years of hard practice to be able to do. Planets and atoms rotate without any appendages or training whatsoever. Is rotating simple? Again, we might say that some cases of rotating are simple or unanalyzable in a metonymic sense if it is done simply by the object in question, given whatever criteria of "doing simply" one might adopt. Rube Goldberg made a lucrative career of inventing machines that do simple actions—that is, actions we usually do simply—in ludicrously complex ways. Psychological capacities too may be derivatively or metonymically simple depending on how the relevant objects do them. It's true that we often think of psychological properties as being complex as types. But we also think that they can be done simply. There are simple decisions and complex ones. Neither is less of a decision. It is compatible with mechanistic explanation of mind that some objects do psychological activities simply. The capacities would be metonymically simple in these cases.

The fact that an activity is done at a lower level of objects does not make the activity type basic in a metaphysical sense, and the fact that an object does something more simply or does fewer things or operates in a simpler environment does not make what it does simple in a metaphysical sense. So when psychological capacities are ascribed to parts, nothing follows about their metaphysical basicness or simplicity.

Both sides can agree that there are metaphysically complex activities with metaphysically simple components, and that there are metaphysically basic and non-basic activities. The difference is that for the Literalist full-blooded psychological capacities may be found in all four of these categories, while for the homuncular functionalist full-blooded psychological capacities must be complex and non-basic. For the Literalist, the capacities are full-blooded wherever they are found, and their metaphysical simplicity or basicness is a further matter we currently know little about. It behooves us to investigate the matter. The point now is that decomposition of psychological capacities is not abandoned in Literalism. What is abandoned is the idea that these decompositions will map onto the mereological hierarchy of objects and *their* relations of basicness and simplicity.

We know independently that there is no simple mapping between metaphysically simple or basic capacities and simple or basic objects. Activity concepts form hierarchies of troponyms, related by manners of doing and other kinds of information (Fellbaum 1990; Fellbaum and Miller 1990; Miller and Fellbaum 1991; Figdor 2017). But troponym hierarchies, and the activities these concepts pick out, are not the same as set-membership hierarchies familiar from objects and object-concepts. Activities do not form the same nested hierarchies. The same activity concept (or activity) can be both a superordinate and a part of another activity: jogging is a manner of moving, but a part of jogging (pushing off with one's foot) is itself also moving. In contrast, a chair is a kind of furniture, but a chair leg is not also furniture. Troponym hierarchies are also much flatter than set-membership hierarchies. We often ascribe the same activities across many object levels because there are more levels in object hierarchies than in activity hierarchies. In short, activity individuation and composition are out of sync with object individuation and composition.

This disconnect in individuation schemes and hierarchies for objects and activities can make mechanistic explanation difficult in cases where there are no objects to which the activities can be assigned. If models and laws organize activities, objects bind them into levels apt for mechanistic explanation. Bechtel (2008: 66–7) provides a nice example of the problem and its solution from the history of chemistry. In chemistry we wanted to explain chemical reactions, such as fermentation. The problem was that "elemental decomposition [into atoms or molecules] was too low a level at which to characterize changes, while decomposing fermentation into fermentations simply invoked the vocabulary designed to explain the overall behavior to describe the operations of its components". The solution was reached when organic chemists "figured out" an intermediate level of functional groups—e.g. hydroxyl groups—that could perform various explanans activities.

In his telling, Bechtel uses the example to illustrate the challenge of identifying the explanans operations. My emphasis is on the fact that explanatory advance was made possible by finding *new objects* to which the operations could be assigned.[11] The actual *operations* posited at the

---

[11] It is hard to overestimate how difficult this is. The modeling literature usually presupposes an appropriately individuated real-world system in order to focus on the relationship between that system and the model that represents it.

new mesoscale level of composition were nothing new: adding and removing. These operations, in these cases, are not basic or simple in either the metonymic or the metaphysical sense. Basic objects don't add and remove, and adding or removing a hydroxyl group is analyzable into activities assigned to more basic objects.

In the case of neural mechanisms, we have not yet found reliable mesoscale levels of objects—what Simon (1962: 471–2) called "stable intermediate forms"—to be the nexuses in a mechanistic explanation of brain-based minds. We also don't know what models will apply at the mesoscales, although the Temporal Difference model discussed in Chapter 3 may be considered a proof of concept. If we find a new mesoscale neural object and its behavior satisfies an old cognitive model, the object and its cognitive capacity may help explain the same or a different cognitive capacity ascribed to the human.

8.5 Concluding Remarks

In this chapter I considered the implications of Literalism for mechanistic explanation of mind, considered as a prime example of the quest for naturalization. The homuncular functionalist framework questions the epistemic legitimacy of ascribing the same psychological capacities to parts and wholes, since it forbids such ascriptions on pain of explanatory inadequacy. I argued that outside of psychology we do not require that a type of activity or capacity never reappear in a part–whole hierarchy as a condition of satisfactory mechanistic explanation. It is false in general that repetition of capacity types at multiple levels is viciously circular, question-begging, idle, or unilluminating, or that a capacity is full-blooded at only one level at which it ascribed. While the mind may be exceptionally poorly understood, I also argued that discharging its spookiness need not be conceived in terms of the gradual exsanguination of mental concepts into non-mental concepts. Models of cognitive capacities can give us the objective perspective we need for discharging without the need for ever-stupider homunculi.

I also argued that Literalism is fully compatible with the idea that the mind is analyzable. It just rejects the homuncular functionalist's implicit assumption that the metaphysical hierarchies of simple/complex or basic/non-basic operations will mirror the mereological decomposition of objects and the familiar levels picture based on object composition. We should not expect to read off from the mereological object hierarchy what the metaphysically basic or simple capacities are.

This need to consider object and activity hierarchies on their own terms carries over to our understanding of the science of psychology. At a certain level of abstraction, psychology is a science that is currently aimed both at a particular level of object composition and at particular types of operational complexity. It involves the study of animals (mainly humans) and their behavior. These two features are being pried apart in the conceptual shift that Literalism draws our attention to.[12] If plants and bacteria behave, they are potentially subjects of psychological science, even if they are not animals or not multicellular. This is just to point out, in somewhat different terms, the contingency of the restriction of psychological predicates to humans in the transition from the Original to the Manifest Image. The restriction was *ipso facto* a restriction to a kind of composite object. This relation between proper psychological predications and a level in an object hierarchy is contingent. As a result, psychology can be a science of certain capacities, or a science of certain levels at which those capacities are manifested. Different psychologists may focus on one or the other, even without a formal institutional split.

In Chapter 9, I consider a reason for rejecting Literalism stemming from moral rather than epistemic concerns. What does Literalism imply for the nature of moral status and corresponding treatment, given that moral status is closely connected to the possession (or not) of cognitive capacities?

---

[12] It follows that the debate over whether psychology is autonomous in some strong sense from neuroscience must be precisified. The debate question might better be put this way: is the study of psychological properties exhibited by humans (and other entities of similar complexity) autonomous from the study of the neural-level properties of these entities? Literalism is neutral on this question, since it does not include a position on what autonomy requires.

9

Literalism and Moral Status

9.1 General Remarks

Literalism holds that scientific uses of psychological predicates across a wide range of unexpected domains are literal with the same reference that they have in the human domain. There is psychological continuity between humans and nonhumans, although which capacities are shared across domains is an empirical matter. Human psychological capacities can still be special, but in the way human vision is special without being the standard for all vision or the best vision. In contrast, anti-Literalists hold that many human capacities are metaphysically exceptional. Human cognition is the standard for genuineness or realness or full-bloodedness. Putative psychological capacities of nonhumans are compared to this standard and found wanting. Nonhumans either have exsanguinated versions of the capacities or don't have them at all.

In this chapter my concern is what psychological continuity may imply for moral status. Having higher (i.e. non-perceptual) psychological capacities and having high moral status are tightly bound together in ethics. Personhood is often defined in terms of higher psychological capacities, such as self-consciousness and rationality, and persons so defined deserve high moral status and the respectful treatment that possessing this status entails. Bennett and Hacker (Bennett et al. 2007: 134) express this standard view when they state that the possession of psychological capacities is what makes a human being a person, which is, "roughly speaking, to possess such abilities as would qualify one for the status of a moral agent". Similarly, McMahan (1996: 31) states that "[h]ow a being ought to be treated depends, to some significant extent, on its intrinsic properties—in particular, its psychological properties and capacities".[1]

---

[1] The link McMahan draws would hold even if psychological properties or capacities were extrinsic (or externalistic).

So if personhood "marks the moral threshold above which equal respect for the intrinsic value of an individual's life is required" (Kittay 2005: 101), and personhood typically depends on possession of psychological properties, does Literalism imply that we owe bacteria, neurons, plants, and so on the same respect accorded to human persons? Does it entail that we must acknowledge the full moral status of many more nonhumans than we do? One might instead argue that since nonhumans do not possess full moral status or deserve the same treatment as humans, then any theory that has these entailments must be false.

Moreover, even if our absolute superiority in moral status is not diminished, it seems that our relative superiority must be if Literalism is true. Yet many find mere comparison with nonhumans offensive. Jaworska and Tannenbaum (2014: 242) acknowledge, in the course of defending their account of moral status, that "some readers may find the comparison of a human being to a dog offensive". In general, animal metaphors are often found offensive and dehumanizing (Haslam et al. 2011). This sense of our superiority is behind McMahan's (1996: 13) comparison involving an anencephalic baby: "Lacking the capacity for consciousness, it has no capacity for well-being at all. It makes no more sense to say an anencephalic is unfortunate, or badly off, than it does to say these things of a plant." Regarding the plant, Bennett and Hacker could not have said it better. A theory that implies that humans are morally comparable to dogs and plants must have gone off the rails somewhere.

In what follows, I'll first set the stage for considering this pragmatic objection to Literalism with a sample of the links between psychological properties and moral status drawn in philosophy and psychology. I then turn to philosophical and empirical research on anthropomorphism and dehumanization to articulate the problem raised for Literalism. These literatures highlight our use of psychological ascriptions as tools to create, maintain, and police social boundaries, including moral boundaries that constrain practices of harming or refraining from harming others. I will argue that fears of radical expansion of moral status or loss of relative superiority in moral status are overblown. There are ways we can avoid radical disruptions in current moral boundaries consistently with Literalism. Since Literalism does not touch our motivations for drawing the boundaries, it is likely that we will avail ourselves of these methods in the short term. However, since Literalism undermines our rational justifications for why we draw the boundaries where we do, it prompts us to reconsider the grounds of moral status from a non-anthropocentric perspective.

9.2 Psychological Ascriptions and Moral Status

The connection between psychological capacity possession and possession of moral status is about as close as that between free will and moral responsibility. "Moral status" is understood here broadly to include moral status as patient or agent, and to admit of degrees. (I will write in terms of agency for brevity, and use "status" and "standing" interchangeably.) In turn, an entity with moral status is entitled to superior treatment than one without it or with lesser status. For example, one might hold, with Bentham and others, that an animal has some moral status because it can suffer, even if it lacks the cognitive capacity to understand moral rules and so lacks full moral status. We care deeply about moral status because of its implications for treatment, and we care deeply about psychological ascriptions because of their implications for moral status.

Here I present an illustrative sample of versions of these connections drawn in philosophy and psychology. Environmental ethicists who attempt to ground intrinsic value in and some moral status to unaltered nature (e.g. Johnson 1993) are notable exceptions to the general agreement about these connections. While Literalism is compatible with the possibility that ecosystems have cognitive capacities, that view goes beyond what I have argued for here (but see Chapter 10). On the other hand, if there are independent grounds of moral status that make psychological properties or the potential for having them sufficient for moral status but not necessary (if they are necessary), Literalism would be silent about such non-psychological grounds.

In philosophy of mind, as noted in Chapter 5, Bennett and Hacker voice a common assimilation of possession of the psychological with personhood and moral standing. Many critically consider this relation in the context of discussing the possible mindedness of machines or animals. For example, Dennett (1978a: 422) notes a close, if not necessary, connection in his discussion of computers and pain. He writes that the "parochiality" of our notion of real pain (as opposed to synthetic pain) is not obviously irrational given its role in "defining our moral community. There can be no denying that our concept of pain is inextricably bound up with (which may mean something less strong than *essentially connected with*) our ethical intuitions, our senses of suffering, obligation, and evil." Regarding intentional states, Newen and Bartels (2007: 283) link interdisciplinary debates over whether animals possess concepts to the drawing of the moral line: "Philosophers' interest in this matter primarily derives from the conviction that concepts are a key factor in distinguishing human beings from non-human animals. This cognitive difference is then exploited to justify important distinctions in the ethical status of human beings as opposed to that of other animals."

The link between psychological capacities and special treatment gets empirical support from psychology. The thought of eating animals whose intelligence is made salient is correlated with greater disgust, whereas the capacities of those who are destined for the platter are downgraded (Bastian et al. 2012; Ruby and Heine 2012). We tend to ascribe higher cognitive capacities to animals we perceive as more similar to ourselves, and, relative to vegetarians, meat-eaters ascribe less psychological and emotional complexity to animals (Bilewicz et al. 2011). In dehumanization (discussed further below), we deny that outgroup members have top-drawer psychological capacities by drawing analogies between them and nonhuman animals or machines. This practice is often a prelude to and justification for causing them harm (Opotow 1990; Leyens et al. 2001; Harris and Fiske 2006; Haslam 2006). Stanley Milgram (1974: 10) noted that participants in his experiments of obedience to authority often devalued the alleged shock victim afterwards as "stupid and stubborn": "these subjects found it necessary to view him as an unworthy individual, whose punishment was made inevitable by his own deficiencies of intellect and character".

Last but not least, ethicists overwhelmingly affirm this close relationship between moral standing and psychological capacities. Some of the specific capacities suggested as being required for full moral status include being able to engage in practical reasoning, being self-aware, being able to value, and being able to care as opposed to merely desire (Jaworska and Tannenbaum 2013). Cognition also grounds differences in degrees of moral status; for example, Jaworska and Tannenbaum (2014: 243) "presume that . . . a sophisticated cognitive capacity grounds your higher moral status" relative to your dog's. The more sophisticated the capacities, the higher the moral status. For the most part, possession of psychological capacities, high moral status, and the right to respectful treatment form a priceless package delivered only to humans, and sometimes only to certain humans. No other (non-divine) beings enjoy this privileged position by default.

The above literatures do not question the idea that healthy adult human cognitive capacities are the standard for full-blooded cognition. Default thresholds for possession of full moral status are set relative to this standard. For this reason the most problematic cases of moral status in ethics are those of humans who do not and never will possess the psychological capacities that meet this standard. The default moral inequivalence between humans and nonhumans must be finessed to justify extending higher moral status to humans who don't meet the psychological criteria and withholding that status from nonhumans who do (or who come closer than the humans do). Some argue that humans with severe psychological impairments have high status anyway (Kittay 2005). Others argue that such humans may have only indirect moral standing or else that nonhuman animals who have equivalent capacities should have the equivalent moral status as a comparable human (McMahan 1996, 2005; Singer 2009).[2]

Literalism affects this moral landscape even if the traditional connection between psychological properties and moral status is unchanged. This is because it questions the human standard for full-bloodedness and our best evidence for determining possession of a psychological capacity. First, it holds that the normal adult human case is not the standard for full-blooded psychological properties. Plants and bacteria can have full-blooded capacities to make decisions, for example, and it is an open question what else they may have. So even if full-blooded capacities and high moral status remain as tightly linked as tradition holds, Literalism implies that many more entities than we have thought may be entitled to high moral status. Merely requiring the possession of sophisticated psychological capacities associated with high functioning adult humans does not rule out extending the highest moral status to nonhumans. The

---

[2] In an indirect duty view, an animal may have a certain treatment status without having the relevant psychological properties because a human who has the properties (and therefore moral standing) would be harmed if the animal were harmed (e.g. Carruthers 1989, for whom any *nonconscious* experiences of animals are not intrinsically qualifying). Similarly, McMahan (1996) argues that we have only indirect moral duties to humans with severe congenital cognitive impairments due to our relations to their caregivers, who have moral standing due to their psychological features.

connection is severed only if a psychological capacity must have human-specific features in order to count as "sophisticated". However, this would be an explicit denial of Literalism rather than a pragmatic rejection of it due to undesirable consequences.

Second, Literalism holds that our best evidence of these full-blooded capacities in other biological entities is not ordinary observation. So it entails that the issue of which things have full moral status is not determined, as it traditionally has been, by the judgments of biologically unsophisticated adults, however philosophically sophisticated they may be. But even if the biologically unsophisticated adult becomes sophisticated, the criteria can be modified to maintain the traditional verdicts. Modification can be effected by claiming that full-blooded capacities must have the human-specific bells and whistles. As noted, this would be an explicit denial of Literalism. Alternatively, one might adjust the criteria for full moral status without explicitly rejecting Literalism by distinguishing more sharply which full-blooded capacities are relevant for high moral status.

I think this latter option is likely in the short run. The long run is a different story. Literalism makes it difficult to continue to justify the superior moral status of humans over nonhumans in the face of scientific advances without resorting to an arbitrary speciesism. Ultimately, it makes the anthropocentrism of traditional judgments about moral status questionable, and perhaps unsustainable.

9.3 The Short Term: Moral Status and Anthropomorphism

The dual perspectives of anthropomorphism and dehumanization help illuminate the role of psychological ascriptions in drawing moral status borders. These perspectives also enable me to show how we can adjust to Literalism with minimal border adjustment.

Dehumanization and anthropomorphism are "inverse" processes (Epley et al. 2007: 864; Haslam et al. 2008b: 55–6). Dehumanization is the denial of traits that we consider either unique to humans or part of human nature or essence (Smith 2001; Haslam 2006; in what follows, I will use "denial" to include ascribing inferior versions of psychological traits). Anthropomorphism is the ascription of human properties, particularly psychological properties, to nonhumans (Caporael 1986: 215; Keeley 2004: 522). In both cases the characteristics ascribed need not be psychological, but I will focus on psychological capacities. In dehumanization, we presuppose that humans have the psychological capacities that are withheld from an individual human or human group. In anthropomorphism, we presuppose that nonhumans don't have the psychological capacities ascribed to an individual nonhuman or nonhuman group.

Epley et al. (2007) argue that we can ascribe a "humanlike" trait to a nonhuman without endorsing the ascription on reflection, let alone presupposing the entity does not really have it.[3] For example, we may ascribe stubbornness to a slow computer without endorsing the ascription, and so presumably without presupposing what it does or doesn't really have. However, classifying a trait as "humanlike" itself indicates a prepotent, if revisable, belief (i.e. presupposition) that it is humans-only, or at least that humans are the standard for the trait. Otherwise it would be difficult to distinguish anthropomorphic ascriptions from non-anthropomorphic ones. An ascription of vision to a dog is not classified as anthropomorphic because we don't presuppose vision to be "humanlike". In contrast, the question of whether an ascription of loyalty to a dog is anthropomorphic arises because loyalty is presupposed to be "humanlike". In other words, one might well interpret being "humanlike" to include vision as well as loyalty, but only the latter ascription is (possibly) anthropomorphic when being "humanlike" is interpreted as being humans-only.[4] So classifying an ascription as anthropomorphic or an ascribed trait as "humanlike" are ways of initially rejecting an Anti-Exceptionalist view of the ascription or the trait. (This in turn suggests that the use of the term for a nonhuman should be interpreted as, e.g.,

---

[3] On their view (Epley et al. 2008: 145; see also Epley et al. 2007), strong anthropomorphism involves both ascribing a "humanlike" property to a nonhuman and endorsing on reflection that ascription as accurate, while weak anthropomorphism involves just ascribing. In those terms, my concern is with strong anthropomorphism. The ascription/endorsement distinction is useful for investigating the factors that predict who is more likely to ascribe a humanlike trait to a nonhuman and when. These include a desire to explain behavior and to satisfy a need for social ties. Presumably the same motivations prompt us to ascribe humanlike traits *to humans*. But if the traits are already considered "humanlike", the ascriptions to humans are presumed non-anthropomorphic.

[4] Thanks to Thomas Butler for suggesting this further clarification of the point. For the record, Wilkes (1975) provides an early defense of a literal interpretation of allegedly anthropomorphic ascriptions.

metaphorical: see Chapter 6.) This presupposition can be overridden upon reflection, in keeping with Epley et al.'s distinction between ascription and endorsement. But it does presume that the Anti-Exceptionalist shoulders the burden of proof for showing that an ascription isn't really anthropomorphic or a trait isn't in fact humans-only. I have already argued that this burden of proof has shifted (see especially Chapter 4), but it will not play a role here.

Both dehumanization and anthropomorphism also have associated practices of differential morally relevant treatment of the dehumanized or anthropomorphized. The literature on dehumanization in particular makes vividly clear that the moral community is a gated community.[5] In dehumanization, we grant and withhold entry into the social and moral community by ascribing (or withholding) specific or any psychological properties to (or from) individuals or groups. We also determine social and moral ranking by drawing distinctions of quality in psychological capacities (e.g. Leyens et al. 2001; Fiske et al. 2002; Haslam et al. 2005; Gray et al. 2007; Loughnan and Haslam 2007; Loughnan et al. 2010; Gray et al. 2011; Jack et al. 2013). We liken enemies to animals that elicit disgust and act from instinct, and demean social inferiors by ascribing inferior psychological capacities to them or less frequently ascribing higher reasoning capacities or social emotions to them. Bezuidenhout's metaphorical description of the toddler in "Our piglet is getting dirty" (see Chapter 6) could easily have been dehumanizing given a different background story. These psychologically-drawn boundaries for in- and outgroup membership and rank serve in turn as a primary justification for differential treatment of people depending on their status. This is especially notable when we want or need to harm them or have already done so. The extremes of harmful treatment are genocide and slavery, but dehumanization also helps reduce inhibitions against treating others less well in subtle ways (Leyens et al. 2001).

---

[5] The dehumanization literature assumes realism regarding psychological ascriptions. The fact that a human has the relevant property explains her membership in the moral community. However, one might hold, in orthodox Intentional Stance-y terms, that *all there is* to having moral status is to be ascribed the relevant psychological properties. For some, this might be reason enough to reject instrumentalism regarding moral status: surely it can't be *that* easy to have high moral status or to be a person.

Although it has not garnered as much attention, anthropomorphism plays a mirror-image social role. Just as dehumanized humans are socially demoted or cast out, "anthropomorphized, nonhuman entities become social entities" (Caporael 1986: 215). When humanlike traits are ascribed to nonhumans, people are more likely to treat the nonhumans "as moral agents worthy of respect and concern [rather than] treated merely as objects" (Epley et al. 2007: 864); "[c]onsciousness, intention, desire, and regret are all the very sorts of humanlike emotions that turn nonhuman agents into *moral* agents" (Epley et al. 2008: 152). However, their moral agency is presumably no more genuine than their possession of the ascribed traits (or than the genuineness of the traits). By classifying the ascription as anthropomorphic, we do not grant the nonhuman agents membership in the moral community even if we treat them in some ways *as if* they belong. Genuine property possession and moral status are both *as if*.[6]

It follows that classifying an ascription as anthropomorphic may not just be a metaphysical decision, and that interpreting a psychological predicate as not being literal may not just be a semantic decision. Both have moral overtones if not also moral grounds. Both practices reinforce social and moral boundaries that keep nonhumans out of the moral community. A. O. Rorty argues that one's deciding that a nonhuman thinks is at the same time one's deciding to treat him in certain favorable ways:

> When I say that someone thinks, I am not only describing *his* behavior. I am also expressing *my* decision to treat him in a certain way: to reason with him rather than to dismantle him, to avoid hurting his feelings wantonly, to treat him with dignity. Someone who denies that slaves, animals, and machines think is expressing a decision not to take the similarities between their behavior and his as relevant to the determination of his actions.   (Rorty 1962: 118)

Observation is insufficient to determine whether a thing *really* thinks, *really* makes ethical decisions, and so on (Rorty 1962: 119–20; Epley et al.

---

[6] This may be relativized to contexts or times. It is inconsistent to think one is anthropomorphizing when using rats as models of human behavior. In the lab, they are model organisms for investigating depression; outside the lab, they are outsiders extraordinaire and frequent sources of dehumanizing analogy. It may be that sciences where psychological ascriptions are accepted as real (non-anthropomorphic) will also be fields in which heretofore human social power relationships will also be theorized (Melinda Fagan, personal communication).

2007: 865; 2008: 144). What counts as a relevant similarity or difference for *real* thinking also reflects a decision about whether we will treat them with the respect due to a moral agent. Whether what counts as relevant is *determined* by this decision about treatment or just expresses the decision, beliefs about whether something has a (real) psychological property and whether it deserves respectful treatment are tightly bound in practice. The decision as to which ascriptions are believed to be anthropomorphic or non-literal is frequently accompanied by, if not metaphysically dependent on, the same types of exclusionary attitudes made explicit in dehumanization.[7]

Given the consequences of drawing moral boundaries and the psychological tools we use to draw them, it is easy to see why we might reject Literalism for pragmatic reasons even if it is the best interpretation of psychological predicates. Nevertheless, we have a number of ways of making compensatory adjustments that would keep existing moral boundaries largely intact consistently with Literalism. Literalism implies radical revision in what we thought were the natural joints of cognition, but it does not entail radical revision in social and moral boundaries.

Anthropomorphic ascriptions are based on induction from our knowledge of human traits and modulated by our knowledge of non-humans (Caporael 1986: 218–19; Epley et al. 2007: 865). As I've argued previously, our best evidential grounds for determining if an entity really has a psychological property (or has a real psychological property) have changed. But the scientific evidence that determines relevant similarity for the purposes of scientific inquiry does not determine the grounds for relevant similarity for the purposes of establishing and maintaining moral status and subsequent treatment. We care enough about the scientific facts to take them into account when supporting

---

[7] Hitchcock and Knobe (2009) and Knobe (2010) have argued that causal judgments and judgments of intentional action are "influenced" by moral judgments, perhaps by way of constraining the range of relevant counterfactual possibilities that we entertain. In superficially similar terms, one could put the view defended here as the claim that judgments of literal vs. non-literal use (in particular, whether we classify an ascription as anthropomorphic or dehumanizing, or neither) are influenced by our interest in drawing morally-relevant social boundaries. These seem to be very different ways in which social and moral concerns can be brought to bear on non-moral similarity judgments, and I see no reason to force them into a single theoretical framework.

our beliefs about which ascriptions are anthropomorphic. We want these beliefs to track the truth. But our interests in social and moral inclusion or exclusion and ranking are far too important to leave to science. We can have our cake and eat it too if the human features that aren't needed for full-blooded capacities are needed for moral status.

Thus, suppose ascriptions of decision-making and preferring to fruit flies and neurons, *inter alia*, come to be considered literal even by the biologically unsophisticated. The same goes for language, of which human language might just be a special case (e.g. Harms 2004: 31), and other prized capacities. Yet suppose we also find it offensive to confer moral status on fruit flies (etc.). If a mathematical model turns out to apply to a type of entity that one did not consider deserving of moral status, one can implicitly tighten up one's criteria of relevant similarity for moral purposes. We can still agree with the Literalist that fruit flies really make decisions, and base this belief entirely on our best scientific evidence. But we can still hold that at least some of the many differences between fruit fly and human decision-making are relevant for the type of decision-making that confers moral status. Empirical and philosophical discoveries do not determine which similarities and differences in behavior of slaves, machines, and monkeys will be considered relevant for decisions about how to treat them (Rorty 1962: 119). In some cases, a capacity that we find to be shared with the undesirable entity might be deprived of moral relevance, in the way vision is considered irrelevant to whether an entity has moral status. (From the current perspective, this disconnect between vision and moral status is not accidental.) In other cases, we might decide that the morally relevant version of the capacity is the one only humans possess. The relevant differences may be those that yield species-specific special cases of the capacities ("The fruit flies have it, but they don't have the relevant version of it"), or they may be the species-specific contexts in which the capacities are deployed ("The fruit flies have it, but they don't or can't exercise it in the relevant contexts").[8] In this way, while nature may fix a best individuation scheme

---

[8] There can also be independent scientific grounds for individuating psychological categories more or less coarsely or finely. I note this because the idea that psychological ascriptions may differ in grain is familiar from the literature on multiple realization (e.g. Kim 1992; Bechtel and Mundale 1999; Figdor 2010): maybe octopi and humans both feel pain, but pain may be (or may not be) a coarse-grained category in such uses. In others, a more fine-grained notion may be invoked. The model of pain discussed in Dennett (1978a)

for the purposes of scientific inquiry (or so we can assume), it needn't fix which individuation scheme will be optimal for our social and moral purposes.

These compensatory adjustments can be done non-consciously. Caporael (1986: 216) notes that contemporary anthropomorphism with respect to cars, computers, and other technologies occurs "despite objective knowledge" (Epley et al. 2007: 866 call this "agent knowledge" or knowledge about humans or oneself). Moreover, our individuation schemes may never be made explicit unless prompted by a request to explain how an object is the same or different (Caporael 1986: 223). If Literalism is correct, we don't yet have much objective knowledge regarding the proper domain of the psychological. We form beliefs about which ascriptions are anthropomorphic from an anthropocentric perspective. Even so, our pragmatic interests can prevent new scientific evidence from proportionately altering these beliefs. A relevant comparison is to how we may continue to believe that members of human outgroups lack top-drawer psychological capacities despite scientific evidence otherwise. We may even seek evidence for those beliefs (pressing our thumbs on the scale all the while). In comparison to the resistance we show towards revising prejudicial beliefs in the light of scientific evidence, neutralizing the impact of Literalism on moral boundaries should be a piece of cake.

In effect, the short-term effect of psychological continuity is likely to parallel that of Greene and Cohen's (2004) assessment of the impact of neuroscience on the law. On their view, if we knew all the neurobiological sources of our actions, our practices of punishing wrongdoers would continue, albeit with non-retributive justification (and perhaps with some adjustments in who gets punished and how). Similarly, psychological continuity promises to change nothing and everything in our practices of social and moral exclusion and ranking. Literalism does not affect our interest in maintaining these social structures or our aversion to downgrading our relative superiority. The crux of the moral issue is not so much whether an entity has certain capacities, but whether we are willing to accept entities of its kind into the moral community. If we don't want to change the current borders, we don't really have to.

might well ground a category that includes many kinds of entities, whether or not we intuitively think they feel pain.

9.4 The Long Term: Moral Status and Anthropocentrism

Nevertheless, this way of protecting existing moral borders is inherently unstable. The class of capacities we can justifiably consider exclusively human is continually shrinking. Further scientific advances may show that the morally relevant versions of the capacities are possessed by more entities than minor border revisions allow in, or that more entities use their capacities in morally relevant contexts. Each border adjustment in response to such advances makes more patent an element of arbitrariness in the way those borders are drawn.[9] The legitimacy of the borders would be severely undermined if we maintained our high moral status by upping the ante each time that status is threatened. Attempts to secure rights for nonhumans would expose this arbitrariness. For example, Stephen Wise (2015: 35–7) of The Nonhuman Rights Project argued in a New York court for a writ of habeas corpus on behalf of two chimpanzees, Hercules and Leo, claiming that chimpanzees "are self-conscious, they have a theory of mind, they can understand what others are thinking, they understand that they are individuals, that they existed yesterday, that they are going to exist tomorrow, that their lives mean something to them" and so on.[10] If we changed the morally relevant versions of these capacities or of the contexts in which they are used whenever it turns out that chimps satisfy the old requirements, the conferral of moral status would become a shell game.

The root of the problem is the anthropocentrism built into the way moral boundaries are currently drawn.[11] Denying and granting moral status to entities dissimilar to the normal adult human (including impaired human infants) requires a lot of fancy footwork, some of which already has the appearance of arbitrary speciesism. Chimpanzees have better working memory than normal adult humans do, at least for numerals (Inoue and Matsuzawa 2007; though see Cook and Wilson 2010), yet don't have moral

[9] The fact that we appear to hold people morally responsible and adjust the metaphysics of free will to suit (e.g. Roskies and Nichols 2008) may be a related phenomenon.

[10] The habeas petition was eventually denied in July 2015, although the judge also determined that a human or corporation (such as the Nonhuman Rights Project) has standing to bring a lawsuit directly on behalf of a nonhuman animal without having to allege injury to a human. <https://www.nonhumanrights.org/hercules-leo/> (accessed Nov. 13, 2017).

[11] Sober (2005: 115) ends with a similar suggestion, albeit without elaboration.

status or do so only tenuously. Humans who lack any working memory at all have moral status nonetheless. An advanced robot that can't feel pain won't have interests, yet a severely impaired human who has congenital inability to feel pain with anhidrosis (CIPA) will. Singer (2009) has argued that we should adopt a graded view of moral status that depends on cognitive ability but applies equally to humans and nonhumans. He distinguishes three main ways to ground superior moral status to all humans above all nonhumans—religious, speciesist, and cognitive—and argues that only the third way, applied in a non-speciesist way, is defensible. But because the standard for cognition is still the human one, even his more ecumenical view remains anthropocentric if not speciesist. This disadvantages nonhumans simply by maintaining the human case as the appropriate standard.

Accounts of moral status which emphasize the social relations in which an entity stands may appear to be on better ground (e.g. Kittay 2005; Jaworska and Tannenbaum 2014). On these views, moral status doesn't depend wholly or primarily on possession of sophisticated psychological capacities or on the potential for possessing them, but rather on the sorts of relationships a being has with other humans. But these accounts too can be anthropocentric in that the human community is the relevant community or the relevant relationships are those formed with humans. The anthropocentricity of these relations is highlighted in debates about human–robot interactions. Even if robots stand in relations to humans that we conventionally associate with rich mental capacities, the ensuing debate has been over whether these relationships are genuine (Sparrow and Sparrow 2006; Turkle 2006). The theme is also explored in popular movies (e.g. *Lars and the Real Girl, Her*). This is the same debate we encountered for psychological properties transposed into the domain of social relationships.

Perhaps moral anthropocentrism, if not speciesism, is defensible on the grounds that moral status is inescapably a human-dependent concept. But this is too quick. As noted, anthropocentric standards can facilitate speciesism because they encourage downgrading other beings' capacities relative to the familiar human case. But they also leave us vulnerable to downgrading ourselves relative to cognitively enhanced beings who adopt the same attitude towards traditional standard human cognitive endowments (Agar 2013; Wasserman 2013). As noted in Chapter 7, we may justify the claim that we have better capacities than nonhumans because we're more flexible in our behavior. We also link this variability to our uniqueness as individuals—a classification with clear moral overtones. So we're in trouble if we make or encounter beings that have more behavioral flexibility than we do. We would be obliged to ascribe even better capacities to them. So even if it weren't arbitrary, speciesism as currently conceived is not worth defending.

Finally, even if the concept of morality has its origins in human social relationships, it does not follow that social relationships won't be found between nonhuman entities. If we find mathematical models of social relationships that apply across domains, we would again be confronted with the problem of how to respond to this evidence of similarity. Indeed, social scientists are busy employing network modeling tools to explore the structure of human social relationships, including political organizations, group formation, and networks of collaboration, influence, novelty, advice, and communication (e.g. Guimera et al. 2005; Baronchelli et al. 2006; Froese et al. 2014). It is highly likely that at least some of these models of social interaction will be successfully extended to interactions between nonhumans (Figdor forthcoming). Plant scientists are already paving the way in their domain by expanding the conceptual repertoire of plant species interactions to include facilitation, not just predation and competition (Bruno et al. 2003).

One might respond that the only sort of morality we care about is the human variety, and our interest in ourselves is not arbitrary. This may be true, but what still needs defense is the idea that drawing the borders of the moral community based on intuitive human standards is justified. The longer-term impact of Literalism lies in how it prompts us to question this traditional anthropocentrism, whether in the form of a human standard for real cognition, the morally-relevant *capacity du jour*, or real social relationships. We may continue to rely on traditional verdicts regarding psychological capacities and social relationships as heuristics for assessing moral status, but such judgments are not clearly justified as robust standards for determining moral status. Avoiding arbitrary speciesism motivates finding grounds for moral status that do not depend on intuitions about human psychological capacities or human social relationships.

9.5 The Fear of Mechanistic Reduction

An independent source of morally-related anxiety about Literalism may come from the idea of reduction. Literalism might be taken to imply that we don't deserve our high status, rather than that nonhumans don't deserve their low status. Immediately after the passage cited in Section 9.3, Rorty writes:

> Conversely, someone who holds that the cognitive and emotive behavior of human beings is not really different in kind from that of a well-programmed and complex machine is committed to treating men as he would treat machines: giving them as much care and consideration as is necessary to get the maximum use from them.   (Rorty 1962: 118)

This quotation is a clear expression of what may seem to follow from psychological continuity. The fear may be most obvious in relation to machines. But fear of being treated like a plant or vegetable—which, by traditional implication, has no moral status or psychological properties—is essentially the same. Many fear that an explanation of mental capacities in terms of brain function will undermine moral behavior by making vivid that we are nothing but biological machines. Biological continuity (evolution) plus physicalism imply at least some degree of psychological continuity, but the question has been whether the continuity spreads upward and inward from basic or peripheral capacities like pain and vision to sophisticated capacities we have considered exclusively ours.

This fear of being stripped of person status within the Scientific Image (acknowledged but not endorsed by Sellars) has become pressing given advances in neuroscience, independently of Literalism. Even if we are not computers, being biological machines is no better. Roskies and Nichols (2008: 377 and fn. 16) summarize the inference as follows:

> Among the neuroethical worries raised by technological advances in neuroscience is that our improving scientific understanding of higher brain functions will cause the public to view currently unexplained psychological phenomena such as choice and decision-making as merely mechanical processes . . . The projected upshot of this potential change of belief is that . . . we will come to realize that we lack free will, and, consequently moral responsibility. This in turn, the argument continues, will undermine the moral fabric of our society; the chaos that will result is left to our imaginations.

The fear is to some extent confirmed in perceptions of medical practice, where doctors are sometimes thought to treat patients as just bags of body parts (Haslam 2006). Freedman (1998: 136) argues against pharmacological cognitive enhancement because it is a way of treating ourselves and responding to others as mere machines: in the enhancement debate, "what is at stake is a conception of ourselves as responsible agents, not machines". Levy (2009: 73) responds, consistently with the connection between psychology and moral status, that "it certainly does not follow from the fact that we are built out of machines that it is appropriate to treat us as machines; we, unlike the machines out of which we are built, are *not* mindless. Human beings live in a space of reasons."[12]

Levy's conclusion is correct, although not because of the reason he gives. Rather, the inference from "Machines and humans both think" or "We are biological machines" to "We will treat each other just as badly we do machines" requires in addition the premise that the natural facts determine the social and moral facts. Rorty makes the same assumption to draw her conclusion.

But for the same reason that the natural facts do not determine moral status, they do not determine how we will treat each other if we take to heart, in the most vivid and unavoidable terms, that we are biological machines. Discoveries in the biological sciences may only change the justifications we give for exclusion, inclusion, or ranking in the moral community. If we want to maintain our exceptional status, we can find ways to do so consistently with physicalism or psychological continuity.

9.6 Concluding Remarks

If humankind has suffered "three humiliations" (Schleim 2012)—the Copernican theory that the Earth is not the center of the universe, the Darwinian theory that humans evolved from common ancestors with apes, and the Freudian theory that we have limited control over and knowledge of our psychological lives—then perhaps psychological continuity is the fourth. If we are not psychological exceptions, it seems to follow either that we do not merit more value than anything that we once thought did not possess those capacities, or else that other things have more value than we once thought they did. Both implications have important consequences for behavior, given that we tend to treat entities that lack or have lesser psychological capacities worse than we treat or expect to be treated ourselves (whether "we" are humans in general or ingroup members).

---

[12] As already noted, the reasons we offer for moral judgment may have nothing to do with their actual causes, which may be morally irrelevant (Haidt 2001; Wheatley and Haidt 2005).

I have argued that the initial inference gets things backwards: we are highly motivated to draw social and moral distinctions between us and them—between humans and nonhumans, and between human ingroups and outgroups—and the classification of an ascription as anthropomorphic (non-literal) is affected by this interest. Our knowledge of the natural facts plays a role, but not a determining one. So while Literalism entails a difference in our knowledge of the natural facts, this does not entail a difference in social and moral boundaries. We have options for adjusting the borders in ways we find most palatable or least offensive. However, these adjustments will be increasingly delegitimized the more they appear designed to restrict high moral status to ourselves in the face of scientific discoveries. In the longer run, Literalism prompts us to reconsider the anthropocentricism of the current grounds of moral status, whether these are psychological capacities or social relationships. There are already independent sources of such pressure in the form of advanced robots filling social roles and technological enhancements of the ordinary cognitive capacities that have long been the benchmark for full moral status.

10

Concluding Summary

In his discussion of the Manifest and Scientific Images, Sellars (1963/
1991: 4) considers the task of the philosopher as follows:

The philosopher is confronted not by one complex many-dimensional picture,
the unity of which, such as it is, he must come to appreciate; but by two pictures
of essentially the same order of complexity, each of which purports to be a
complete picture of man-in-the-world, and which, after separate scrutiny, he
must fuse into one vision.

I hope to have accomplished at least part of this task in this book. My first
main aim was to show that there is an urgent need to understand the
increasing use of psychological terms to characterize new scientific
discoveries and the impact of new scientific methods on our reasons
for endorsing such predications as factive. These empirical, theoretical,
and methodological developments throughout biology are at odds with
received philosophical and folk perspectives on the proper domain of
psychological predicates and how we determine that proper domain.

My second main aim was to show that the most plausible view of
psychological predicates is Literalism, which holds that the terms are
used literally with the same reference across human and nonhuman
domains. We are finally in a position to see how conceptual change in
psychology might come about and what the Scientific Image (or a fused
Manifest-Scientific Image) might actually look like. Literalism agrees that
psychological concepts were anchored in human experience and allows
that we can still gain insight from phenomenology. But these concepts
were never defined within a closed formal system. We should not expect
them to remain untouched by advances in scientific knowledge.

Literalism is not the only response to these developments, but I have
argued that the other views are much worse off. The Nonsense view
fails on its own terms. The Metaphor view fails to provide independent,

non-question-begging evidence that the uses are metaphorical. Both variants of the Technical view fail to justify the distinction in reference between humans and nonhumans that they assert. Literalism, in contrast, is easily able to account for the entire range of uses and the epistemic reasons for the extensions. It shows that scientists are using these terms rationally in ways that are appropriately constrained by the empirical evidence. It embraces the facts that the main goal of science is to discover what the natural world is like and the main goal of peer-reviewed scientific publications is to express those discoveries in fact-stating discourse.

I have also considered two implications of Literalism that some might find unacceptable. It conflicts with the mechanistic explanation of mind as articulated in the homuncular functionalist view of psychological explanation. It also appears to imply that current borders and ranks of moral status must be radically altered to allow in far more kinds of nonhumans or diminish our relative superiority over nonhumans. I've argued that homuncular functionalism is not epistemically justified and that discharging the mind does not require us to posit homunculi at all. Rather than being in conflict with naturalism, Literalism prompts us to reconsider what is required for a naturalistic explanation of mind. I also argued that adjustments to the current distinctions of moral status can be significantly attenuated, given our strong interest in maintaining them and the associated ways of treating others and being treated ourselves. In the long run, however, Literalism prompts us to reconsider the anthropocentric way in which we now conceptualize the grounds of moral status. This anthropocentrism is inherited from the anthropocentrism of our current standards for real psychological properties.

Literalism also offers further benefits across disciplines that stem from its new perspective on the nature of psychological predicates. In particular, it promotes theoretical and empirical inquiry in a number of active areas of research.

First, inferences from humans to nonhumans and vice versa are both enhanced by a non-anthropocentric view of the mind. The questions of whether something has a cognitive capacity and whether it is intuitively like us are clearly distinguished. Key evidence of similarity of cognitive capacity is via models applied across domains, while similarity to human behavior may explain why we more readily recognize a capacity in a nonhuman or accept them as having moral status. Evidence relevant to the first question can also be derived directly from a wider range of model organisms, with greater epistemic security, and with less room for conceptual confusion. This was demonstrated by DasGupta et al.'s (2014) study involving fruit flies, the relevance of which to human genetically-based cognitive disorders would be obscured if we could not interpret the flies as making decisions. It also promotes innovative theorizing by prompting us to take initial model-based similarities as grounds for sameness of reference across domains that we may have not previously considered importantly similar at all.

Second, research in animal cognition stands to benefit. Literalism furthers the methodological shift in cognitive and behavioral ethology to observing animals in the wild (or in more normal habitats) rather than in highly artificial contexts. This change was motivated by the idea that animal behavior is what animals do in their environments, not what they do in ours. Literalism motivates in addition the separation of our reasons for inferring to a nonhuman species' capacity from our judgments of whether their behavior is more or less similar (and more or less inferior) to ours. We have begun to stop thinking that animals must act just like humans to be intelligent. Literalism implies that we should also stop thinking that intelligence should be measured by our intuitive human standard.

Third, Literalism has obvious application to debates about psycho-logical ascriptions to social groups, artificial intelligent agents, and even divinities—for example, whether such ascriptions to God should be understood univocally (Duns Scotus) or else equivocally or analogically (Thomas Aquinas).[1] Literalism undermines the idea that real cognition must be brain-based or headbound, even within the realm of biology. It is not an argument for extended or group cognition so much as a new perspective on certain presuppositions in those debates. Literalism offers a new way to understand ascriptions to groups or extended individuals and to assess evidence for or against such ascriptions. Finally, research involving mathematical models of social behavior is expanding rapidly. These models promise to raise questions about ascriptions of social concepts parallel to those discussed in this book regarding psychological concepts. In general, Literalism implies a different perspective in any

---

[1] Thanks to Georg Theiner for pointing out the relevance to divine predication.

debate, philosophical or otherwise, where a necessary or very close connection between the human and the really (or full-bloodedly) cognitive is drawn or presupposed. This will include debates about the nature of agency, social interaction, intentional action, and the self.

Fourth, a lingering Cartesian internalism about the causes of our behavior may begin to fade away for good. Many of the intuitive or initial philosophical responses to Literalism show that the importance of external factors for understanding our behavior has not really sunk in. It is still true that we are different, but so is every species. Literalism implies that many important cognitive differences may correspond to the different ways capacities are expressed in different contexts, at different timescales, by different bodies. Psychological research suggests that we should not continue to expect that the best explanation of our behavior will be largely in terms of uniquely human capacities inside our heads. Literalism suggests that we should also not continue to assume that the best explanation of nonhuman behavior is largely in terms of external factors. Literalism implies greater convergence in the background assumptions regarding how we should explain the behavior of humans and nonhumans.

Literalism is not a scientistic position, at least not as I conceive of it. There is more to human life than what science tells us about ourselves. But science is also telling us that there is more to nonhuman life than what we have long assumed. As I see it, Literalism suggests that humanistic inquiry is not, and should not be, just about humans.

Bibliography

Agar, N. (2013). Why is it Possible to Enhance Moral Status and Why Doing So is Wrong? *Journal of Medical Ethics* 39(2): 67–74.

Allen, C. and M. Bekoff (1999). *Species of Mind: The Philosophy and Biology of Cognitive Ethology*. Cambridge, MA: MIT Press.

Alon, U. (2007). Network Motifs: Theory and Experimental Approaches. *Nature Reviews Genetics* 8: 450–61.

Alpi, A., M. Amrhein, A. Bertl, M. Blatt, E. Blumwald, F. Cervone, J. Dainty, M. DeMichaelis, E. Epstein, A. Galston, M. Goldsmith, C. Hawes, R. Hell, A. Hetherington, H. Hofte, G. Guergens, C. Leaver, A. Moroni, A. Murphy, K. Oparka, P. Perata, H. Quader, T. Rausch, C. Ritzenthaler, A. Rivetta, D. Robinson, D. Sanders, B. Scheres, K. Schumacher, H. Sentenac, C. Slayman, C. Soave, C. Somerville, L. Taiz, G. Thiel, and R. Wagner (2007). Plant Neurobiology: No Brain, No Gain? *Trends in Plant Sciences* 12(7): 135–6.

Anderson, M. (2010). Neural Reuse: A Fundamental Organizational Principle of the Brain. *Behavioral and Brain Sciences* 33: 245–313.

Andrews, K. (2009). Politics or Metaphysics? On Attributing Psychological Properties to Animals. *Biology and Philosophy* 24(1): 51–63.

Andrews, K. (2012). *Do Apes Read Minds? Toward a New Folk Psychology*. Cambridge, MA: MIT Press.

Aral, S. and D. Walker (2012). Identifying Influential and Susceptible Members of Social Networks. *Science* 337: 337–41.

Ariew, R. and D. Garber, ed. and trans. (1989). *Leibniz: Philosophical Essays*. Indianapolis and Cambridge: Hackett.

Aristotle (1983). *Physics Books I and II*. Trans. with Introduction and Notes by W. Charlton. Oxford: Clarendon Press.

Aronson, E. and J. Mills (1959). The Effect of Severity of Initiation on Liking for a Group. *Journal of Abnormal and Social Psychology* 59: 177–81.

Arora, V. and G. Boer (2006). Simulating Competition and Coexistence Between Plant Functional Types in a Dynamic Vegetation Model. *Earth Interactions* 10: Paper No. 10.

Attneave, W. (1961). In Defense of Homunculi. In Walter A. Rosenblith, ed., *Sensory Communication*. Cambridge, MA: MIT Press, 777–81.

Auletta, G. (2011). *Cognitive Biology: Dealing with Information from Bacteria to Minds*. New York: Oxford University Press.

Bach, K. (1994). Conversational Implicature. *Mind & Language* 9: 124–62.

Baluška, F. (2009). *Plant–Environment Interactions: From Sensory Plant Biology to Active Plant Behavior.* Berlin and Heidelberg: Springer-Verlag.

Baluška, F. and S. Mancuso (2007). Plant Neurobiology as a Paradigm Shift not only in the Plant Sciences. *Plant Signaling & Behavior* 2(4): 205–7.

Baluška, F. and S. Mancuso (2009). Plants and Animals: Convergent Evolution in Action? In F. Baluška, ed., *Plant–Environment Interactions: From Sensory Plant Biology to Active Plant Behavior.* Berlin and Heidelberg: Springer-Verlag, 285–301.

Baluška, F., F. Volkmann, and S. Mancuso (2006). *Communication in Plants: Neuronal Aspects of Plant Life.* Berlin and Heidelberg: Springer-Verlag.

Baronchelli, A., M. Felici, V. Loreto, E. Caglioti, and L. Steels (2006). Sharp Transition Towards Shared Vocabulary in Multi-Agent Systems. arXiv:physics/0509075v2.

Baronchelli, A., R. Ferrer-i-Cancho, R. Pastor-Satorras, N. Chater, and M. Christiansen (2013). Networks in Cognitive Science. *Trends in Cognitive Sciences* 17(7): 348–60.

Barrett, J., R. Richert, and A. Driesenga (2001). God's Beliefs vs. Mother's: The Development of Nonhuman Agent Concepts. *Child Development* 72(1): 50–65.

Barsalou, L. (1987). The Instability of Graded Structure: Implications for the Nature of Concepts. In U. Neisser, ed., *Concepts and Conceptual Development: Ecological and Intellectual Factors in Categorization.* Cambridge: Cambridge University Press, 101–40.

Barsalou, L. (1993). Flexibility, Structure, and Linguistic Vagary in Concepts: Manifestations of a Compositional System of Perceptual Concepts. In A. Collins, S. Gathercole, M. Conway, and P. Morris, eds., *Theories of Memory.* Hove: Lawrence Erlbaum Associates, 29–101.

Barsalou, L. (1999). Perceptual Symbol Systems. *Behavioral and Brain Sciences* 22: 577–660.

Barto, A. (1995). Adaptive Critics and the Basal Ganglia. In J. Houk, J. Davis, and D. Beiser, eds., *Models of Information Processing in the Basal Ganglia.* Cambridge, MA: MIT Press, 215–32.

Barto, A., M. Mirolli, and G. Baldassarre (2013). Novelty or Surprise? *Frontiers in Psychology* 4 (article 907). <http://dx.doi.org/10.3389/fpsyg.2013.00907>.

Bassler, B. (2002). Small Talk: Cell-to-Cell Communication in Bacteria. *Cell* 109: 421–4.

Bastian, B., S. Loughnan, N. Haslam, and H. Radka (2012). Don't Mind Meat? The Denial of Mind to Animals Used for Human Consumption. *Personality and Social Psychology Bulletin* 38(2): 247–56.

Batterman, R. and C. Rice (2014). Minimal Model Explanations. *Philosophy of Science* 81(3): 349–76.

Bechtel, W. (2008). *Mental Mechanisms*. New York and Abingdon: Lawrence Erlbaum Associates.

Bechtel, W. (2009). Constructing a Philosophy of Science for Cognitive Science. *Topics in Cognitive Science* 1: 548–69.

Bechtel, W. and A. Abrahamsen (2005). Explanation: A Mechanist Alternative. *Studies in History and Philosophy of Biological and Biomedical Sciences* 36: 421–41.

Bechtel, W. and J. Mundale (1999). Multiple Realizability Revisited: Linking Cognitive and Neural States. *Philosophy of Science* 66(2): 175–207.

Bechtel, W. and R. Richardson (1993). *Discovering Complexity*. Princeton, NJ: Princeton University Press.

Beer, R. (2004). Autopoiesis and Cognition in the Game of Life. *Artificial Life* 10: 309–26.

Beer, R. (2015). Characterizing Autopoiesis in the Game of Life. *Artificial Life* 21: 1–19.

Bekoff, M., C. Allen, and G. Burkhardt, eds. (2002). *The Cognitive Animal: Empirical and Theoretical Perspectives on Animal Cognition*. Cambridge, MA: MIT Press.

Benham, B. (2009). Analogies and Other Minds. *Informal Logic* 29(2): 198–214.

Ben-Jacob, E., I. Becker, Y. Shapira, and H. Levine (2004). Bacterial Linguistic Communication and Social Intelligence. *Trends in Microbiology* 12(8): 366–72.

Ben-Jacob, E., D. Coffey, and H. Levine (2012). Bacterial Survival Strategies Suggest Rethinking Cancer Cooperativity. *Trends in Microbiology* 20(9): 403–10.

Ben-Jacob, E., Y. Shapira, and A. Tauber (2006). Seeking the Foundations of Cognition in Bacteria: From Schrödinger's Negative Entropy to Latent Information. *Physica A* 359: 495–524.

Bennett, M., D. Dennett, P. M. S. Hacker, and J. Searle (2007). *Neuroscience & Philosophy: Brain, Mind, and Language*. New York: Columbia University Press.

Bennett, M. R. and P. M. S. Hacker (2003). *Philosophical Foundations of Neuroscience*. Malden, MA and Oxford: Blackwell.

Bentley, R. A., M. O'Brien, and W. Brock (2014). Mapping Collective Behavior in the Big-Data Era. *Behavioral and Brain Sciences* 37: 63–119.

Berent, I. (2013). The Phonological Mind. *Trends in Cognitive Sciences* 17(7): 319–27.

Berridge, K. (2004). Motivation Concepts in Behavioral Neuroscience. *Physiology & Behavior* 81: 179–209.

Bezuidenhout, A. (2001). Metaphor and What is Said: A Defense of a Direct Expression View of Metaphor. *Midwest Studies in Philosophy* 25: 156–86.

Bezuidenhout, A. (2002). Truth-Conditional Pragmatics. *Philosophical Perspectives* 16: 105–34.

Bibikov, S., A. Miller, K. Gosink, and J. Parkinson (2004). Methylation-Independent Aerotaxis Mediated by the *Escherichia coli* Aer Protein. *Journal of Bacteriology* 186(12): 3730-7.

Bilewicz, M., R. Imhoff, and M. Drogosz (2011). The Humanity of What We Eat: Conceptions of Human Uniqueness among Vegetarians and Carnivores. *European Journal of Social Psychology* 41: 201-9.

Bitterman, M., R. Menzel, A. Fietz, and S. Schäfer (1983). Classical Conditioning of Probiscus Extension in Honeybees (*Apis mellifera*). *Journal of Comparative Psychology* 97(2): 107-19.

Blakemore, C. (1977). *Mechanics of the Mind*. Cambridge: Cambridge University Press.

Boisvert, M. and D. Sherry (2006). Interval Timing by an Invertebrate, the Bumble Bee *Bombus impatiens*. *Current Biology* 16(16): 1636-40.

Bose, I. and R. Karmakar (2003). Simple Models of Plant Learning and Memory. *Physica Scripta* T106: 9-12.

Bose, J. (1926). *The Nervous Mechanism of Plants*. London and New York: Longmans, Green and Co.

Bourret, R. and A. Stock (2002). Molecular Information Processing: Lessons from Bacterial Chemotaxis. *Journal of Biological Chemistry* 277(12): 9625-8.

Bowdle, B. and D. Gentner (2005). The Career of Metaphor. *Psychological Review* 112(1): 193-216.

Boyd, R. (1993). Metaphor and Theory Change: What is "Metaphor" a Metaphor For? In A. Ortony, ed., *Metaphor and Thought*, 2nd edn. Cambridge: Cambridge University Press, 481-532.

Brandom, R. (1994). *Making It Explicit: Reasoning, Representing, and Discursive Commitment*. Cambridge, MA: Harvard University Press.

Brandom, R. (1995). Knowledge and the Social Articulation of the Space of Reasons. *Philosophy and Phenomenological Research* 55(4): 895-908.

Brandom, R. (1997). Précis of Making It Explicit. *Philosophy and Phenomenological Research* 57(1): 153-6.

Brandom, R. (2010). Conceptual Content and Discursive Practice. *Grazer Philosophische Studien* 81: 13-35.

Brenner, E. D., R. Stahlberg, S. Mancuso, F. Baluška, and E. Von Volkenburgh (2007). Reply to Alpi et al.: Plant Neurobiology: The Gain is More than the Name. *Trends in Plant Sciences* 12(7): 285-6.

Brenner, E. D., R. Stahlberg, S. Mancuso, J. Vivanco, F. Baluška, and E. Von Volkenburgh (2006). Plant Neurobiology: An Integrated View of Plant Signaling. *Trends in Plant Sciences* 11(8): 413-19.

Brown, S. (2008). Polysemy in the Mental Lexicon. *Colorado Research in Linguistics* 21(1): 1-12.

Bruno, J., J. Stachowicz, and M. Bertness (2003). Inclusion of Facilitation into Ecological Theory. *Trends in Ecology and Evolution* 18(3): 119–25.

Bullmore, E. and O. Sporns (2009). Complex Brain Networks: Graph and Theoretical Analysis of Structural and Functional Systems. *Nature Reviews Neuroscience* 10: 186–98.

Bullmore, E. and O. Sporns (2012). The Economy of Brain Network Organization. *Nature Reviews Neuroscience* 13: 336–49.

Burnham, J., S. Collart, and B. Highison (1981). Entrapment and Lysis of the Cyanobacterium *Phoridium luridum* by Aqueous Colonies of *Myxococcus Xanthus* PCO2. *Archives of Microbiology* 129(4): 285–94.

Cacioppo, J., G. Berntson, and H. Nusbaum (2008). Neuroimaging as a New Tool in the Toolbox of Psychological Science. *Current Directions in Psychological Science* 17(2): 62–7.

Callender, C. and J. Cohen (2006). There is No Special Problem about Scientific Representation. *Theoria* 55: 67–85.

Calvo, P., F. Baluška, and A. Sims (2016). "Feature Detection" vs. "Predictive Coding" Models of Plant Behavior. *Frontiers in Psychology* 7 (Article 1505).

Calvo Garzon, F. (2007). The Quest for Cognition in Plant Neurobiology. *Plant Signaling & Behavior* 2(4): 208–11.

Calvo Garzon, F. and F. Keijzer (2009). Cognition in Plants. In F. Baluška, ed., *Plant– Environment Interactions: From Sensory Plant Biology to Active Plant Behavior*. Berlin and Heidelberg: Springer-Verlag, 247–66.

Camerer, C. (2008). Neuroeconomics: Opening the Gray Box. *Neuron* 60: 416–19.

Camerer, C., G. Loewenstein, and D. Prelec (2005). Neuroeconomics: How Neuroscience Can Inform Economics. *Journal of Economic Literature* 43: 9–64.

Camp, E. (2006). Metaphor in Mind: The Cognition of Metaphor. *Philosophy Compass* 1(2): 154–70.

Caporael, L. (1986). Anthropomorphism and Mecanomorphism: Two Faces of the Human Machine. *Computers in Human Behavior* 2: 215–34.

Caramazza, A. (1986). On Drawing Inferences about the Structure of Normal Cognitive Systems from the Analysis of Patterns of Impaired Performance: The Case for Single-Patient Studies. *Brain and Cognition* 5: 41–66.

Carandini, M. (2012). From Circuits to Behavior: A Bridge Too Far? *Nature Neuroscience* 15(4): 507–9.

Carey, S. (1988). Conceptual Differences between Children and Adults. *Mind & Language* 3: 167–81.

Carruthers, P. (1989). Brute Experience. *Journal of Philosophy* 86(5): 258–69.

Carston, R. (2002). *Thoughts and Utterances: The Pragmatics of Explicit Communication*. Oxford: Blackwell.

Carston, R. (2010). Metaphor: Ad hoc Concepts, Literal Meaning, and Mental Images. *Proceedings of the Aristotelian Society* 110(3): 297–323.

Carston, R. (2012). Metaphor and the Literal/Nonliteral Distinction. In K. Allan and K. M. Jaszczolt, eds., *The Cambridge Handbook of Pragmatics*. Cambridge: Cambridge University Press, 469–92.

Cartwright, N. (1983). *How the Laws of Physics Lie*. Oxford: Clarendon Press.

Chalmers, D. (1995). Facing Up to the Problem of Consciousness. *Journal of Consciousness Studies* 2(3): 200–19.

Chang, H. (2012). The Persistence of the Everyday in the Scientific. *Philosophy of Science* 79(5): 690–700.

Chemero, A. (2009). *Radical Embodied Cognitive Science*. Cambridge, MA and London: MIT Press.

Chemero, A. and M. Silberstein (2008). After the Philosophy of Mind: Replacing Scholasticism with Science. *Philosophy of Science* 75(1): 1–27.

Chen, E., P. Widick, and A. Chatterjee (2008). Functional-Anatomical Organization of Predicate Metaphor Processing. *Brain & Language* 107: 194–202.

Chen, M. K., V. Lakshminarayana, and L. Santos (2006). How Basic are Behavioral Biases? Evidence from Capuchin Monkey Trading Behavior. *Journal of Political Economy* 114(3): 517–37.

Cherniak, C. (1986). *Minimal Rationality*. Cambridge, MA: MIT Press.

Chirimuuta, M. (2014). Minimal Models and Canonical Neural Computations: The Distinctness of Explanation in Neuroscience. *Synthese* 191(2): 127–53.

Churchland, P. M. (1970). The Logical Character of Action-Explanations. *Philosophical Review* 79(2): 214–36.

Churchland, P. M. (1981). Eliminative Materialism and the Propositional Attitudes. *Journal of Philosophy* 78(2): 67–90.

Churchland, P. S. and T. Sejnowski (1988). Perspectives on Cognitive Neuroscience. *Science*, New Series 242: 741–5.

Clark, A. (2008). *Supersizing the Mind: Embodiment, Action, and Cognitive Extension*. Oxford: Oxford University Press.

Clark, A. (2013). Whatever Next? Predictive Brains, Situated Agents, and the Future of Cognitive Science. *Behavioral and Brain Sciences* 36(3): 181–253.

Clark, A. and D. Chalmers (1998). The Extended Mind. *Analysis* 58(1): 7–19.

Colby, C. (1991). The Neuroanatomy and Neurophysiology of Attention. *Journal of Child Neurology* 6: S90–S118.

Cook, P. and M. Wilson (2010). In Practice, Chimp Memory Study Flawed. *Science* 328: 1228.

Craver, C. (2001). Role Functions, Mechanisms, and Hierarchy. *Philosophy of Science* 68(1): 53–74.

Craver, C. (2002). Structures of Scientific Theories. In P. Machamer and M. Silberstein, eds., *Blackwell Guide to the Philosophy of Science*. Oxford: Blackwell, 55–79.

Craver, C. (2007). *Explaining the Brain: Mechanisms and the Mosaic Unity of Neuroscience*. Oxford: Oxford University Press.

Crick, F. (1995). *The Astonishing Hypothesis*. London: Touchstone Books.

Crick, F. and C. Koch (1990). Toward a Neurobiological Theory of Consciousness. *Seminars in the Neurosciences* 2: 263–75.

Crist, E. (2000). *Images of Animals: Anthropomorphism and Animal Mind*. Philadelphia: Temple University Press.

Croft, W. and D. A. Cruse (2004). *Cognitive Linguistics*. Cambridge: Cambridge University Press.

Cummins, R. (1975). Functional Analysis. *Journal of Philosophy* 72(20): 741–65.

Cummins, R. (1983). *The Nature of Psychological Explanation*. Cambridge, MA and London: MIT Press.

Cummins, R., P. Poirier, and M. Roth (2004). Epistemological Strata and Rules of Right Reason. *Synthese* 141: 287–331.

Damasio, A. and D. Tranel (1993). Nouns and Verbs are Retrieved with Differently Distributed Neural Systems. *Proceedings of the National Academy of the Sciences USA* 90: 4957–60.

Darwin, C. (1880). *The Power of Movement in Plants*. London: John Murray. <http://darwin-online.org.uk/content/frameset?pageseq=1&itemID=F1325&viewtype=text>.

Dasgupta, S., C. Ferreira, and G. Miesenböck (2014). FoxP Influences the Speed and Accuracy of a Perceptual Decision in *Drosophila*. *Science* 344: 901–4.

Davidson, D. (1979/2001). What Metaphors Mean. In S. Sacks, ed., *On Metaphor*. Chicago: Chicago University Press. Reprinted in D. Davidson, *Inquiries into Truth and Interpretation*. Oxford: Clarendon Press, 245–64.

Davidson, D. (1984). Truth and Meaning. In D. Davidson, *Inquiries into Truth and Interpretation*. Oxford: Clarendon Press, 17–36.

Deneve, S. (2008). Bayesian Spiking Neurons I: Inference. *Neural Computation* 20: 91–117.

Dennett, D. (1969). *Content and Consciousness*. New York: Humanities Press.

Dennett, D. (1975a). Why the Law of Effect Will Not Go Away. *Journal of the Theory of Social Behavior* 5: 169–87. Reprinted in Dennett 1978b, 71–89 (page numbers refer to 1978b).

Dennett, D. (1975b). Three Kinds of Intentional Psychology. In R. Healey, ed., *Reduction, Time, and Reality: Studies in the Philosophy of the Natural Sciences*. Cambridge: Cambridge University Press, 37–61.

Dennett, D. (1978a). Why You Can't Make a Computer That Feels Pain. *Synthese* 38(3): 415–56.

Dennett, D. (1978b). *Brainstorms*. Cambridge, MA: MIT Press.

Dennett, D. (1981/1997). True Believers: The Intentional Strategy and Why it Works. In A. Heath, ed., *Scientific Explanation: Papers based on Herber Spencer*

Lectures given in the University of Oxford. Oxford: Clarendon Press, 150–67. Reprinted in J. Haugeland, ed., *Mind Design II.* Cambridge, MA: MIT Press, 1997, 57–79.

Dennett, D. (1984). *Elbow Room.* Oxford: Oxford University Press.

Dennett, D. (1987). *The Intentional Stance.* Cambridge, MA: MIT Press.

Dennett, D. (1991). Real Patterns. *Journal of Philosophy* 88(1): 27–51.

Dennett, D. (2007). Philosophy as Naïve Anthropology. In M. Bennett, D. Dennett, P. M. S. Hacker, and J. Searle, *Neuroscience & Philosophy: Brain, Mind, and Language.* New York: Columbia University Press, 73–95 (cited as Bennett et al. 2007).

Dennett, D. (2009). Intentional Systems Theory. In A. Beckermann, B. McLaughlin, and S. Walter, eds., *The Oxford Handbook of Philosophy of Mind.* Oxford: Oxford University Press, 339–50.

Dennett, D. (2010). The Evolution of 'Why'. In B. Weiss and J. Wanderer, eds., *Reading Brandom: On Making It Explicit.* London: Routledge, 48–62.

Dennett, D. (2013). Expecting Ourselves to Expect: The Bayesian Brain as Projector. *Behavioral and Brain Sciences* 36: 209–10.

Dennett, D. (2014). The Evolution of Reasons. In B. Bashour and H. Muller, eds., *Contemporary Philosophical Naturalism and its Implications.* New York: Routledge, 47–62.

Dretske, F. (1988). *Explaining Behavior: Reasons in a World of Causes.* Cambridge, MA: MIT Press.

Eldar, E., R. Rutledge, R. Dolan, and Y. Niv (2016). Mood as Representation of Momentum. *Trends in Cognitive Sciences* 20(1): 15–24.

Emery, N. and N. Clayton (2004). The Mentality of Crows: Convergent Evolution of Intelligence in Corvids and Apes. *Science* 306: 1903–7.

Epley, N., A. Waytz, S. Akalis, and J. Cacioppo (2008). When We Need a Human: Motivational Determinants of Anthropomorphism. *Social Cognition* 26(2): 143–55.

Epley, N., A. Waytz, and J. Cacioppo (2007). On Seeing Human: A Three-Factor Theory of Anthropomorphism. *Psychological Review* 114(4): 864–86.

Fagan, M. (2016). Stem Cells and Systems Models: Clashing Views of Explanation. *Synthese* 193(3): 873–907.

Falik, O., P. Reides, M. Gersani, and A. Novoplansky (2003). Self/Non-Self Discrimination in Roots. *Journal of Ecology* 91: 525–31.

Fellbaum, C. (1990). English Verbs as a Semantic Net. *International Journal of Lexicography* 3(4): 278–301.

Fellbaum, C. and G. Miller (1990). Folk Psychology or Semantic Entailment? A Reply to Rips and Conrad (1989). *Psychological Review* 97(4): 565–70.

Field, H. (1973). Theory Change and the Indeterminacy of Reference. *Journal of Philosophy* 70(14): 462–81.

Figdor, C. (2010). Neuroscience and the Multiple Realization of Cognitive Functions. *Philosophy of Science* 77(3): 419–56.

Figdor, C. (2011). Semantics and Metaphysics in Informatics: Towards an Ontology of Tasks. *Topics in Cognitive Science* 3: 222–6.

Figdor, C. (2013). What is the "Cognitive" in Cognitive Neuroscience? *Neuroethics* 6: 105–14.

Figdor, C. (2014). Verbs and Minds. In M. Sprevak and J. Kallestrup, eds., *New Waves in Philosophy of Mind*. Basingstoke: Palgrave Macmillan, 38–53.

Figdor, C. (2017). On the Proper Domain of Psychological Predicates. *Synthese* 194: 4289–310.

Figdor, C. (forthcoming). Big Data and Changing Concepts of the Human. *European Review*.

Fillmore, C. and B. Atkins (2000). Describing Polysemy: The Case of 'Crawl'. In C. Leacock, ed., *Polysemy: Theoretical and Computational Approaches*. Oxford: Oxford University Press, 91–110.

Firn, R. (2004). Plant Intelligence: An Alternative Point of View. *Annals of Botany* 93: 345–51.

Fiske, S., A. Cuddy, P. Glick, and J. Xu (2002). A Model of (often Mixed) Stereotype Content: Competence and Warmth Respectively Follow from Perceived Status and Competition. *Journal of Personality and Social Psychology* 82(6): 878–902.

Fodor, J. (1968a). The Appeal to Tacit Knowledge in Psychological Explanation. *Journal of Philosophy* 65(20): 627–40.

Fodor, J. (1968b). *Psychological Explanation*. New York and Toronto: Random House.

Fodor, J. (1975). *The Language of Thought*. Cambridge, MA: Harvard University Press.

Fodor, J. (1987). *Psychosemantics: The Problem of Meaning in the Philosophy of Mind*. Cambridge, MA: MIT Press.

Fontenot, M., S. Watson, K. Roberts, and R. Miller (2007). Effects of Food Preferences on Token Exchange and Behavioral Responses to Inequality in Tufted Capuchin Monkeys, *Cebus apella*. *Animal Behavior* 74: 487–96.

Freedman, C. (1998). Aspirin for the Mind? Some Ethical Worries about Psychopharmacology. In E. Parens, ed., *Enhancing Human Traits: Ethical and Social Implications*. Washington, DC: Georgetown University Press, 135–50.

French, S. and J. Ladyman (1999). Reinflating the Semantic Approach. *International Studies in the Philosophy of Science* 13(2): 103–21.

Frigg, R. (2006). Scientific Representation and the Semantic View of Theories. *Theoria* 55: 49–65.

Frigg, R. (2010). Models and Fiction. *Synthese* 172: 251–68.

Friston, K. (2010). The Free-Energy Principle: A Unified Brain Theory? *Nature Reviews Neuroscience* 11: 127–38.

Friston, K. (2013). Active Inference and Free Energy. *Behavioral and Brain Sciences* 36(3): 212–13.

Froese, T., C. Gershenson, and L. Manzanilla (2014). Can Government Be Self-Organized? A Mathematical Model of the Collective Social Organization of Ancient Teotihuacan, Central Mexico. *PLoS One* 9(10): e109966.

Galton, A. and R. Mizoguchi (2009). The Water Falls but the Waterfall Does Not Fall: New Perspectives on Objects, Processes and Events. *Applied Ontology* 4(2): 71–107.

Garson, J. (2002). The Introduction of Information in Neurobiology. *Philosophy of Science* 70(5): 926–36.

Gentner, D. (1981). Some Interesting Differences between Nouns and Verbs. *Cognition and Brain Theory* 4(2): 161–78.

Gentner, D. (1982). Why Nouns Are Learned Before Verbs: Linguistic Relativity versus Natural Partitioning. In S. Kuczaj, ed., *Language Development: Language, Cognition, and Culture*. Hillsdale, NJ: Lawrence Erlbaum, 301–34.

Gentner, D. (1983). Structure Mapping: A Theoretical Framework for Analogy. *Cognitive Science* 7: 155–70.

Gentner, D. and L. Boroditsky (2001). Individuation, Relativity, and Early Word Learning. In M. Bowerman and S. Levinson, eds., *Language Acquisition and Conceptual Development*. Cambridge: Cambridge University Press, 215–56.

Gentner, D., B. Bowdle, P. Wolff, and C. Boronat (2001). Metaphor is Like Analogy. In K. Holyoak and B. Kokinov, eds., *The Analogical Mind*. Cambridge, MA: MIT Press, 199–253.

Gentner, D. and M. Jeziorski (1993). The Shift from Metaphor to Analogy in Western Science. In A. Ortony, ed., *Metaphor and Thought*, 2nd edn. Cambridge: Cambridge University Press, 447–80.

Gentner, D. and A. Markman (1997). Structure Mapping in Analogy and Similarity. *American Psychologist* 52(1): 45–56.

Gibbs, R. (1990). Comprehending Figurative Referential Descriptions. *Journal of Experimental Psychology: Learning, Memory, and Cognition* 16(1): 56–66.

Gibbs, R. and M. Tendahl (2006). Cognitive Effort and Effects in Metaphor Comprehension: Relevance Theory and Psycholinguistics. *Mind & Language* 21(3): 379–403.

Giere, R. (1988). *Explaining Science: A Cognitive Approach*. Chicago: University of Chicago Press.

Giere, R. (2004). How Models are Used to Represent Reality. *Philosophy of Science* 71 (suppl.): S742–S752.

Giurfa, M. and R. Menzel (2003). Human Spatial Representation Derived from a Honeybee Compass. *Trends in Cognitive Sciences* 7(2): 59–60.

Glennan, S. (2002). Rethinking Mechanistic Explanation. *Philosophy of Science* 69 (suppl.): S342–S353.

Gleitman, L., K. Cassidy, A. Papafragou, R. Nappa, and J. Trueswell (2005). Hard Words. *Journal of Language Learning and Development* 1: 23–64.

Glock, H.-J. (2009). Can Animals Act for Reasons? *Inquiry* 52(3): 232–54.

Glock, H.-J. (2010). Can Animals Judge? *Dialectica* 64(1): 11–33.

Glucksberg, S. (2001). *Understanding Figurative Language: From Metaphors to Idioms.* Oxford: Oxford University Press.

Glucksberg, S. and B. Keysar (1990). Understanding Metaphorical Comparisons: Beyond Similarity. *Psychological Review* 97(1): 3–18.

Glucksberg, S. and B. Keysar (1993). How Metaphors Work. In A. Ortony, ed., *Metaphor and Thought*, 2nd edn. Cambridge: Cambridge University Press, 401–24.

Glüer, K. and P. Pagin (2012). General Terms and Relational Modality. *Noûs* 46 (1): 159–99.

Godfrey-Smith, P. (2005). Folk Psychology as a Model. *Philosophers' Imprint* 5(6): 1–16.

Godfrey-Smith, P. (2006). The Strategy of Model-Based Science. *Biology and Philosophy* 21: 725–40.

Goodwin, R. (1967). A Growth Cycle. In C. Feinstein, ed., *Socialism, Capitalism, and Economic Growth.* Cambridge: Cambridge University Press, 54–8.

Gorman, J. (2013). Considering the Humanity of Non-Humans. *The New York Times* (Dec. 3). <http://www.nytimes.com/2013/12/10/science/considering-the-humanity-of-nonhumans.html?_r=0>.

Goto, Y., S. Otani, and A. Grace (2007). The Yin and Yang of Dopamine Release: A New Perspective. *Neuropharmacology* 53: 583–7.

Goy, M., M. Springer, and J. Adler (1977). Sensory Transduction in *Escherischia coli*: Role of a Protein Methylation Reaction in Sensory Adaptation. *Proceedings of the National Academy of Sciences of the USA* 74(11): 4964–8.

Gray, H., K. Gray, and D. Wegner (2007). Dimensions of Mind Perception. *Science* 315: 619.

Gray, K., J. Knobe, M. Sheskin, P. Bloom, and L. Barrett (2011). More than a Body: Mind Perception and the Nature of Objectification. *Journal of Personality and Social Psychology* 101(6): 1207–20.

Greene, J. and J. Cohen (2004). For the Law, Neuroscience Changes Everything and Nothing. *Philosophical Transactions of the Royal Society: Biological Sciences* 359(1451): 1775–85.

Greenwood, J. (1999). Understanding the "Cognitive Revolution" in Psychology. *Journal of the History of the Behavioral Sciences* 35(1): 1–22.

Grice, P. (1975/1989). Logic and Conversation. In P. Cole and J. Morgan, eds., *Syntax & Semantics 3: Speech Acts.* New York: Academic Press, 41–58. Reprinted as Ch. 2 in P. Grice, *Studies in the Ways of Words.* Cambridge, MA: Harvard University Press, 1989.

Guimera, R., B. Uzzi, J. Spiro, and L. Nunes Amaral (2005). Team Assembly Mechanisms Determine Collaboration Network Structure and Team Performance. *Science* 308: 697–702.

Gunderson, K. (1969). Cybernetics and Mind–Body Problems. *Inquiry* 12: 406–19.

Haidt, J. (2001). The Emotional Dog and its Rational Tail: A Social Intuitionist Approach to Moral Judgment. *Psychological Review* 108: 814–34.

Halvorson, H. (2012). What Scientific Theories Could Not Be. *Philosophy of Science* 79: 183–206.

Harms, W. (2004). Primitive Content, Translation, and the Emergence of Meaning in Animal Communication. In D. K. Oller and U. Griebel, eds., *Evolution of Communication Systems: A Comparative Approach*. Cambridge, MA: MIT Press, 31–48.

Harris, L. and S. Fiske (2006). Dehumanizing the Lowest of the Low: Neuroimaging Responses to Extreme Outgroups. *Psychological Science* 17: 847–53.

Haslam, N. (2006). Dehumanization: An Integrative Review. *Journal of Personality and Social Psychology* 10(3): 252–64.

Haslam, N., P. Bain, L. Douge, M. Lee, and B. Bastian (2005). More Human than You: Attributing Humanness to Oneself and Others. *Journal of Personality and Social Psychology* 89: 973–50.

Haslam, N., Y. Kashima, S. Loughnan, J. Shi, and C. Suitner (2008a). Subhuman, Inhuman, and Superhuman: Contrasting Humans with Non-Humans in Three Cultures. *Social Cognition* 26(2): 248–58.

Haslam, N. and S. Loughnan (2013). Dehumanization and Infrahumanization. *Annual Review of Psychology* 65: 399–423.

Haslam, N., S. Loughnan, Y. Kashima, and P. Bain (2008b). Attributing and Denying Humanness to Others. *European Journal of Social Psychology* 19: 55–85.

Haslam, N., S. Loughnan, and P. Sun (2011). Beastly: What Makes Animal Metaphors Offensive? *Journal of Language and Social Psychology* 30: 311–25.

Haspelmath, M. (2012). How to Compare Major Word-Classes across the World's Languages. *UCLA Working Papers in Linguistics* 17(16): 109–30.

Haugeland, J. (1993). Pattern and Being. In B. Dahlbom, ed., *Dennett and his Critics*. Oxford: Blackwell, 53–69.

Hazelbauer, G., J. Falke, and J. Parkinson (2008). Bacterial Chemoreceptors: High-Performance Signaling in Networked Arrays. *Trends in Biochemical Sciences* 33(1): 9–19.

Healy, K. (unpublished). *Fuck Nuance*. Presented at the 2015 American Sociological Association meeting. <http://kieranhealy.org/files/papers/fuck-nuance.pdf>.

Hellingwurf, K. (2005). Bacterial Observations: A Rudimentary Form of Intelligence? *Trends in Microbiology* 13(4): 152–8.

Hempel, C. and P. Oppenheim (1948). Studies in the Logic of Explanation. *Philosophy of Science* 15: 135–75.

Hendry, R. and S. Psillos (2006). How to Do Things With Theories: An Interactive View of Language and Models in Science. In J. Brzeziński, A. Klawiter, T. A. F. Kuipers, K. Łastowski, K. Paprzycka, and P. Przybysz, eds., *The Courage of Doing Philosophy: Essays Dedicated to Leszek Nowak*. Amsterdam/New York: Rodopi, 59–115.

Hesse, M. (1966). *Models and Analogies in Science*. Notre Dame, IN: University of Notre Dame Press.

Hesselmann, G., S. Sadaghiani, K. Friston, and A. Kleinschmidt (2010). Predictive Coding or Evidence Accumulation? False Inference and Neural Fluctuations. *PLoS One* 5(3): e9926.

Hitchcock, C. and J. Knobe (2009). Cause and Norm. *Journal of Philosophy* 106 (11): 587–612.

Hodgkin, A. and A. Huxley (1952). A Quantitative Description of Membrane Current and Its Application to Conduction and Excitation in Nerve. *Journal of Physiology* 117: 500–44.

Hohwy, J. (2013). *The Predictive Mind*. Oxford: Oxford University Press.

Hornsby, J. (2000). Personal and Subpersonal: A Defense of Dennett's Early Distinction. *Philosophical Explorations: An International Journal for the Philosophy of Mind and Action* 3(1): 6–24.

Hubel, D. and T. Wiesel (1962). Receptive Fields, Binocular Interaction and Functional Architecture in the Cat's Visual Cortex. *Journal of Physiology* 160: 106–54.

Huebner, B. (2014). *Macrocognition: A Theory of Distributed Minds and Collective Intentionality*. New York: Oxford University Press.

Humphreys, P. (2002). Computational Models. *Philosophy of Science* 69: S1–S11.

Hurley, S. (2003). Animal Action in the Space of Reasons. *Mind & Language* 18(3): 231–56.

Illes, J., M. Moser, J. McCormick, E. Racine, S. Blakeslee, A. Caplan, E. Check Hayden, J. Ingram, T. Lohwater, P. McKnight, C. Nicholson, A. Phillips, K. Sauvé, E. Snell, and S. Weiss (2010). Neurotalk: Improving the Communication of Neuroscience Research. *Nature Reviews Neuroscience* 11: 61–9.

Inagaki, K. and G. Hatano (1987). Young Children's Spontaneous Personification as Analogy. *Child Development* 58: 1013–20.

Inagaki, K. and G. Hatano (1993). Young Children's Understanding of the Mind–Body Distinction. *Child Development* 64: 1534–49.

Inagaki, K. and G. Hatano (1996). Young Children's Recognition of Commonalities between Animals and Plants. *Child Development* 67: 2823–40.

Inoue, S. and T. Matsuzawa (2007). Working Memory of Numerals in Chimpanzees. *Current Biology* 17(23): R1004–R1005.

Irvine, E. (2016). Model-Based Theorizing in Cognitive Neuroscience. *British Journal for the Philosophy of Science* 67: 143–68.

Izhikevich, E. (2007). *Dynamical Systems in Neuroscience: The Geometry of Excitability and Bursting*. Cambridge, MA and London: MIT Press.

Izuma, K. and R. Adolphs (2013). Social Manipulation of Preference in the Human Brain. *Neuron* 78: 563–73.

Jack, A., A. Dawson, and M. Norr (2013). Seeing Human: Distinct and Overlapping Neural Signatures Associated with Two Forms of Dehumanization. *NeuroImage* 79: 313–28.

Jackendoff, R. (1983). *Semantics and Cognition*. Cambridge, MA: MIT Press.

Jackendoff, R. (1990). *Semantic Structures*. Cambridge, MA: MIT Press.

Jackendoff, R. and D. Aaron (1991). Review of G. Lakoff and M. Turner, *More Than Cool Reason: A Field Guide to Poetic Metaphor*. *Language* 67: 320–38.

Jacob, P. (2011). Meaning, Intentionality, and Communication. In C. Maienborn, K. Heusinger, and P. Portner, eds., *Semantics: An International Handbook of Natural Language Meaning vol. 1*. Berlin: De Gruyter/Mouton, 11–24. Manuscript version. <http://www.scalab.cnrs.fr/CNCC09/HandbookJacob.pdf>.

Jaworska, A. and J. Tannenbaum (2013). The Grounds of Moral Status. In E. Zalta, ed., *The Stanford Encyclopedia of Philosophy* (Summer 2013 edition). <http://plato.stanford.edu/archives/sum2013/entries/grounds-moral-status/>.

Jaworska, A. and J. Tannenbaum (2014). Person-Rearing Relationships as a Key to Higher Moral Status. *Ethics* 124(2): 242–71.

Johnson, C. (1997). Metaphor vs. Conflation in the Acquisition of Polysemy: The Case of *See*. In M. K. Hiraga, C. Sinha, and S. Wilcox, eds., *Cultural, Psychological and Typological Issues in Cognitive Linguistics*. Amsterdam: John Benjamins, 155–69.

Johnson, L. (1993). *A Morally Deep World: An Essay on Moral Significance and Environmental Ethics*. Cambridge: Cambridge University Press.

Kahneman, D. (2011). *Thinking, Fast and Slow*. New York: Farrar, Straus & Giroux.

Kaplan, D. and C. Craver (2011). The Explanatory Force of Dynamical and Mathematical Models in Neuroscience: A Mechanistic Perspective. *Philosophy of Science* 78(4): 601–27.

Kapur, S. (2003). Psychosis as a State of Aberrant Salience: A Framework Linking Biology, Phenomenology, and Pharmacology in Schizophrenia. *American Journal of Psychiatry* 160: 13–23.

Keeley, B. (2004). Anthropomorphism, Primatomorphism, Mammalomorphism: Understanding Cross-Species Comparisons. *Biology and Philosophy* 19: 521–40.

Keeton, W. (1967). *Biological Science*. New York: Norton.

Keijzer, F. (2013). The Sphex Story: How the Cognitive Sciences Keep Repeating an Old and Questionable Anecdote. *Philosophical Psychology* 26(4): 502–9.

Keller, E. F. (1983). *Feeling for the Organism: The Life and Work of Barbara McClintock*. New York: W. H. Freeman.

Kelly, C. (1992). Resource Choice in *Cuscuta europaea*. *Proceedings of the National Academy of Sciences USA* 89: 12194–7.

Kelso, J. A. S. (1995). *Dynamic Patterns: The Self-Organization of Brain and Behavior*. Cambridge, MA: MIT Press.

Kim, J. (1992). Multiple Realization and the Metaphysics of Reduction. *Philosophy and Phenomenological Research* 52(1): 1–26.

Kittay, E. (2005). At the Margins of Moral Personhood. *Ethics* 116: 100–31.

Knobe, J. (2010). Person as Scientist, Person as Moralist. *Behavioral and Brain Sciences* 33: 315–65.

Knowles, J. (2002). Is Folk Psychology Different? *Erkenntnis* 57: 199–230.

Knutson, B. and S. Greer (2008). Anticipatory Affect: Neural Correlates and Consequences for Choice. *Philosophical Transactions of the Royal Society: Biological Sciences* 363: 3771–86.

Koch, C. (2012). *Consciousness: Confessions of a Romantic Reductionist*. Cambridge, MA: MIT Press.

Kuhn, T. (1962). *The Structure of Scientific Revolutions*. Chicago: University of Chicago Press.

Kuhn, T. (1993). Metaphor in Science. In A. Ortony, ed., *Metaphor and Thought*, 2nd edn. Cambridge: Cambridge University Press, 533–42.

Lakoff, G. and M. Johnson (1980). *Metaphors We Live By*. Chicago: University of Chicago Press.

Lakoff, G. and M. Turner (1989). *More Than Cool Reason: A Field Guide to Poetic Metaphor*. Chicago: University of Chicago Press.

Lam, H.-M., J. Chiu, M.-H. Hsieh, L. Meisel, I. Oliveira, M. Shin, and G. Coruzzi (1998). Glutamate-Receptor Genes in Plants. *Nature* 396: 125–6.

Langlois, J., J. Ritter, L. Roggman, and L. Vaughn (1991). Facial Diversity and Infant Preferences for Attractive Faces. *Developmental Psychology* 27(1): 79–84.

Laub, M. (2011). The Role of Two-Component Signal Transduction Systems in Bacterial Stress Responses. In G. Storz and R. Hengge, eds., *Bacterial Stress Responses*. Washington, DC: ASM Press, 45–58.

Lenartowicz, A., D. Kalar, E. Congdon, and R. Poldrack (2010). Towards an Ontology of Cognitive Control. *Topics in Cognitive Science* 2: 678–92.

Levin, B. (1993). *English Verb Classes and Alternations*. Chicago: University of Chicago Press.

Levy, A. (2015). Modeling Without Models. *Philosophical Studies* 172: 781–98.

Levy, A. and W. Bechtel (2013). Abstraction and the Organization of Mechanisms. *Philosophy of Science* 80(2): 241–61.

Levy, A. and A. Currie (2015). Model Organisms are Not (Theoretical) Models. *British Journal for the Philosophy of Science* 66(2): 327–48.

Levy, N. (2009). Neuroethics: Ethics and the Sciences of the Mind. *Philosophy Compass* 4(1): 69–81.

Levy, N. (2014). Is Neurolaw Conceptually Confused? *Journal of Ethics* 18: 171–85.

Lewis, D. (1970). How to Define Theoretical Terms. *Journal of Philosophy* 67: 427–46.

Lewis, D. (1994). Reduction of Mind. In S. Guttenplan, ed., *A Companion to Philosophy of Mind*. Oxford: Blackwell, 412–31.

Leyens, J. P., A. Rodriguez-Perez, R. Rodriguez-Torres, R. Gaunt, M. P. Paladino, J. Vaes, and S. Demoulin (2001). Psychological Essentialism and the Differential Attribution of Uniquely Human Emotions to Ingroups and Outgroups. *European Journal of Social Psychology* 31: 395–411.

Loughnan, S. and N. Haslam (2007). Animals and Androids: Implicit Associations between Social Categories and Nonhumans. *Psychological Science* 18(2): 116–21.

Loughnan, S., N. Haslam, T. Murnane, J. Vaes, C. Reynolds, and C. Suitner (2010). Objectification Leads to Depersonalization: The Denial of Mind and Moral Concern to Objectified Others. *European Journal of Social Psychology* 40: 709–17.

Ludvig, E., R. Sutton, and E. J. Kehoe (2012). Evaluating the TD Model of Classical Conditioning. *Learning and Behavior* 40: 305–19.

Lycan, W. (1981). Form, Function, and Feel. *Journal of Philosophy* 78(1): 24–50.

Lycan, W. (1987). *Consciousness*. Cambridge, MA: MIT Press.

Lycan, W. (1988). *Judgment and Justification*. Cambridge: Cambridge University Press.

Lycan, W. (1991). Homuncular Functionalism Meets PDP. In W. Ramsey, S. Stich, and D. Rumelhart, eds., *Philosophy and Connectionist Theory*. Hillsdale, NJ: Lawrence Erlbaum Associates, 259–86.

Lyon, P. (2006). The Biogenic Approach to Cognition. *Cognitive Processing* 7: 11–29.

McClintock, B. (1984). The Significance of Responses of the Genome to Challenge. *Science*, New Series 226: 792–801.

McDowell, J. (1994). *Mind and World*. Cambridge, MA: Harvard University Press.

McDowell, J. (1996). Précis of *Mind and World*. *Philosophical Issues* 7: 231–9.

McDowell, J. (2010). Tyler Burge on Disjunctivism. *Philosophical Explorations* 13: 243–55.

Machamer, P., L. Darden, and C. Craver (2000). Thinking about Mechanisms. *Philosophy of Science* 67(1): 1–25.

McMahan, J. (1996). Cognitive Disability, Misfortune, and Justice. *Philosophy & Public Affairs* 25(1): 3–35.

McMahan, J. (2005). Our Fellow Creatures. *Journal of Ethics* 9: 353–80.

Mahall, B. and R. Callaway (1996). Effects of Regional Origin and Genotype on Intraspecific Root Communication in the Desert Shrub *Ambrosia dumosa* (Asteraceae). *American Journal of Botany* 83(1): 93–8.

Maibom, H. (2003). The Mindreader and the Scientist. *Mind & Language* 18(3): 296–315.

Maienschein, J. (2017). Epigenesis and Preformationism. In E. Zalta, ed., *The Stanford Encyclopedia of Philosophy* (Spring 2017 edition). <https://plato.stanford.edu/archives/spr2017/entries/epigenesis/>.

Margolis, J. (1980). The Trouble with Homuncular Theories. *Philosophy of Science* 47(2): 244–59.

Maturana, H. and F. Varela (1980). *Autopoiesis and Cognition: The Realization of the Living*. Dordrecht: D. Reidel.

Medin, D., R. Goldstone, and D. Gentner (1993). Respects for Similarity. *Psychological Review* 100(2): 254–78.

Mendoza, E., J. Colomb, J. Rybak, H.-J. Pflüger, T. Zars, C. Scharff, and B. Brembs (2014). *Drosophila* FoxP Mutants are Deficient in Operant Self-Learning. *PLoS One* 9(6): e100648.

Milgram, S. (1974). *Obedience to Authority: An Experimental View*. New York: Harper & Row.

Miller, G. A. (2010). Mistreating Psychology in the Decades of the Brain. *Perspectives on Psychological Science* 5: 716–43.

Miller, G. A. and C. Fellbaum (1991). Semantic Networks of English. *Cognition* 41: 197–229.

Miller, G. A., E. Galanter, and K. Pribram (1960). *Plans and the Structure of Behavior*. New York: Holt, Rinehart & Winston.

Miller, R., R. Barnet, and N. Grahame (1995). Assessment of the Rescorla–Wagner Model. *Psychological Bulletin* 117(3): 363–86.

Millikan, R. (1984). *Language, Thought, and Other Biological Categories*. Cambridge, MA: MIT Press.

Millikan, R. (1989). Biosemantics. *Journal of Philosophy* 86(6): 281–97.

Millstein, R., R. Skipper, and M. Dietrich (2009). (Mis)Interpreting Mathematical Models: Drift as a Physical Process. *Philosophy and Theory in Biology* 1: e002.

Milosavljevic, M., J. Malmaud, A. Huth, C. Koch, and A. Rangel (2010). The Drift Diffusion Model Can Account for the Accuracy and Reaction Time of Value-Based Choices under High and Low Time Pressure. *Judgment and Decision Making* 5(6): 437–49.

Mitchell, A., G. Romano, B. Groisman, A. Yona, E. Dekel, M. Kupiec, O. Dahan, and Y. Pilpel (2009). Adaptive Prediction of Environmental Changes by Microorganisms. *Nature* 460: 220–5.

Montague, P., P. Dayan, and T. Sejnowski (1996). A Framework for Mesencephalic Dopamine Systems based on Predictive Hebbian Learning. *Journal of Neuroscience* 16: 1936–47.

Morewedge, C., J. Preston, and D. Wegner (2007). Timescale Bias in the Attribution of Mind. *Journal of Personality and Social Psychology* 93(1): 1–11.

Morgan, M. and M. Morrison (1999). Models as Mediating Instruments. In M. Morgan and M. Morrison, eds., *Models as Mediators*. Cambridge: Cambridge University Press, 10–37.

Mori, M. (1970). The Uncanny Valley. *Energy* 7(4): 33–5.

Morrison, M. (2015). *Reconstructing Reality: Models, Mathematics, and Simulations*. New York: Oxford University Press.

Murphy, G. (1996). On Metaphoric Representation. *Cognition* 60: 173–204.

Nagel, E. (1961). *The Structure of Science*. New York: Harcourt, Brace & World.

Nagel, T. (1974). What Is It Like to Be a Bat? *Philosophical Review* 83(4): 435–50.

Nakagaki, T., H. Yamada, and Á. Tóth (2000). Maze-Solving by an Amoeboid Organism. *Nature* 407: 470.

Neisser, U. (1967). *Cognitive Psychology*. New York: Appleton Century Crofts.

Nersessian, N. (1992). How do Scientists Think? Capturing the Dynamics of Conceptual Change in Science. In R. Giere, ed., *Cognitive Models of Science*. Minneapolis: University of Minnesota Press, 3–44.

Nersessian, N. (2008). *Creating Scientific Concepts*. Cambridge, MA: MIT Press.

Newell, A., J. Shaw, and H. Simon (1958). Elements of a Theory of Problem Solving. *Psychological Review* 65(3): 151–66.

Newen, A. and A. Bartels (2007). Animal Minds and the Possession of Concepts. *Philosophical Psychology* 20(3): 283–308.

Oizumi, M., L. Albantakis, and G. Tononi (2014). From the Phenomenology to the Mechanisms of Consciousness: Integrated Information Theory 3.0. *PLoS Computational Biology* 10(5): e1003588.

O'Keefe, J. and J. Dostrovsky (1971). The Hippocampus as a Spatial Map: Preliminary Evidence from Unit Activity in the Freely-Moving Rat. *Brain Research* 34: 171–5.

Opotow, S. (1990). Moral Exclusion and Injustice: An Introduction. *Journal of Social Issues* 46: 1–20.

Ortony, A., ed. (1993). *Metaphor and Thought*, 2nd edn. Cambridge: Cambridge University Press.

Orzack, S. and E. Sober (1993). A Critical Assessment of Levins's 'The Strategy of Model-Building in Population Biology' (1966). *The Quarterly Review of Biology* 68: 533–46.

Palmer, J., A. Huk, and M. Shadlen (2005). The Effect of Stimulus Strength on the Speed and Accuracy of a Perceptual Decision. *Journal of Vision* 5: 376–404.

Papafragou, A. (1998). Modality and the Semantics–Pragmatics Interface. University College London, dissertation.

Pargetter, R. (1984). The Scientific Inference to Other Minds. *Australasian Journal of Philosophy* 62(2): 158–63.

Pauwels, E. (2013). Mind the Metaphor. *Nature* 500: 523–4.

Pendlebury, M. (1998). Intentionality and Normativity. *South African Journal of Philosophy* 17(2): 142–51.

Penn, D. and D. Povinelli (2013). The Comparative Delusion: The 'Behavioristic'/'Mentalistic' Dichotomy in Comparative Theory of Mind Research. In J. Metcalfe and H. Terrace, eds., *Agency and Joint Attention*. Oxford: Oxford University Press, 62–81.

Pettet, G., D. McElwain, and J. Norbury (2000). Lotka–Volterra Equations with Chemotaxis: Walls, Barriers and Travelling Waves. *IMA Journal of Mathematics Applied in Medicine and Biology* 27: 395–413.

Piccinini, G. and C. Craver (2011). Integrating Psychology and Neuroscience: Functional Analyses as Mechanism Sketches. *Synthese* 183: 283–311.

Platt, M. and P. Glimcher (1999). Neural Correlates of Decision Variables in Parietal Cortex. *Nature* 400: 233–8.

Poldrack, R. (2010). Mapping Mental Function to Brain Structure: How Can Cognitive Neuroimaging Succeed? *Perspectives on Psychological Science* 5(6): 753–61.

Poldrack, R. and T. Yarkoni (2016). From Brain Maps to Cognitive Ontologies: Informatics and the Search for Semantic Structure. *Annual Reviews of Psychology* 67: 20.1–20.26.

Price, C. and K. Friston (2005). Functional Ontologies for Cognition: The Systematic Definition of Structure and Function. *Cognitive Neuropsychology* 22(3): 262–75.

Prinz, J. (2002). *Furnishing the Mind: Concepts and their Perceptual Basis*. Cambridge, MA: MIT Press.

Pylyshyn, Z. (1993). Metaphorical Imprecision and the "Top-Down" Research Strategy. In A. Ortony, ed., *Metaphor and Thought*, 2nd edn. Cambridge: Cambridge University Press, 543–58.

Rakova, M. (2003). *The Extent of the Literal: Metaphor, Polysemy and Theories of Concepts*. Basingstoke and New York: Palgrave Macmillan.

Ramsey, W. (2007). *Representation Reconsidered*. Cambridge: Cambridge University Press.

Ramsey, W. (2016). Untangling Two Questions about Mental Representation. *New Ideas in Psychology* 40: 3–12.

Rankin, C. (2004). Invertebrate Learning: What Can't a Worm Learn? *Current Biology* 14: R617–18.

Rao, R. (2005). Bayesian Inference and Attentional Modulation in the Visual Cortex. *NeuroReport* 16(16): 1843–8.

Rao, R. and D. Ballard (1999). Predictive Coding in the Visual Cortex: A Functional Interpretation of some Extra-Classical Receptive Field Effects. *Nature Neuroscience* 2: 79–87.

Ratcliff, R. (1978). A Theory of Memory Retrieval. *Psychological Review* 85(2): 59–108.

Ratcliff, R. and G. McKoon (2008). The Drift Decision Model: Theory and Data for Two-Choice Decision Tasks. *Neural Computation* 20(4): 873–922.

Ratcliff, R. and P. Smith (2004). A Comparison of Sequential Sampling Models for Two-Choice Reaction Time. *Psychological Review* 111(2): 333–67.

Ratcliff, R., P. Smith, S. Brown, and G. McKoon (2016). Diffusion Decision Models: Current Issues and History. *Trends in Cognitive Sciences* 20(4): 260–81.

Ravenscroft, I. (2010). Folk Psychology as a Theory. In E. Zalta, ed., *The Stanford Encyclopedia of Philosophy* (Fall 2010 edition). <http://plato.stanford.edu/arch ives/fall2010/entries/folkpsych-theory/>.

Recanati, F. (1993). *Direct Reference: From Language to Thought.* Oxford: Blackwell.

Recanati, F. (2001). Literal/Nonliteral. *Midwest Studies in Philosophy* 25: 264–74.

Recanati, F. (2010). *Truth-Conditional Pragmatics.* Oxford: Oxford University Press.

Rescorla, M. (2015). Bayesian Perceptual Psychology. In M. Matthen, ed., *The Oxford Handbook of Philosophy of Perception.* Oxford: Oxford University Press, 694–716.

Rescorla, R. (1988). Pavlovian Conditioning: It's Not What You Think It Is. *American Psychologist* 43(3): 151–60.

Rescorla, R. and A. Wagner (1972). A Theory of Pavlovian Conditioning: Variations in the Effectiveness of Reinforcement and Nonreinforcement. In A. Black and W. Prokasy, eds., *Classical Conditioning II: Current Research and Theory.* New York: Appleton Century Crofts, 64–99.

Reuveny, E. (2013). Ion Channel Twists to Open. *Nature* 498: 182–3.

Rips, L. J. and F. G. Conrad (1990). Parts of Activities: Reply to Fellbaum and Miller (1990). *Psychological Review* 97(4): 571–5.

Robinson, T. and K. Berridge (2000). The Psychology and Neurobiology of Addiction: An Incentive-Sensitization View. *Addiction* 95 (Suppl. 2): S91–S117.

Rolls, E. T., A. S. Browning, K. Inoue, and I. Hernadi (2005). Novel Visual Stimuli Activate a Population of Neurons in Primate Orbitofrontal Cortex. *Neurobiology of Learning and Memory* 84: 111–23.

Rorty, A. O. (1962). Slaves and Machines. *Analysis* 22(5): 118–20.

Rosch, E. (1973). Natural Categories. *Cognitive Psychology* 4: 328–50.

Rosch, E. (1975). Cognitive Representations of Semantic Categories. *Journal of Experimental Psychology* 104(3): 192–233.

Rosch, E. and C. B. Mervis (1975). Family Resemblances: Studies in the Internal Structure of Categories. *Cognitive Psychology* 7: 573–605.

Rosch, E., C. B. Mervis, W. D. Gray, D. M. Johnson, and P. Boyes-Braem (1976). Basic Objects in Natural Categories. *Cognitive Psychology* 8: 382–439.

Roskies, A. (2010). How Does Neuroscience Affect our Concept of Volition? *Annual Reviews of Neuroscience* 33: 109–30.

Roskies, A. and S. Nichols (2008). Bringing Moral Responsibility Down to Earth. *Journal of Philosophy* 105(7): 371–88.

Roth, M. (2013). Folk Psychology as Science. *Synthese* 190: 3971–82.

Ruby, M. B. and S. J. Heine (2012). Too Close to Home: Factors Predicting Meat Avoidance. *Appetite* 59: 47–52.

Ryle, G. (1949). *The Concept of Mind*. Chicago: University of Chicago Press.

Saigusa, T., A. Teto, T. Nakagaki, and Y. Kuramoto (2008). Amoebae Anticipate Periodic Events. *Physical Review Letters* 100 (018101).

Samuelson, L., G. Jenkins, and J. Spencer (2015). Grounding Cognitive Level Processes in Behavior: The View from Dynamic Systems Theory. *Topics in Cognitive Science* 7: 191–205.

Sanchez, A. and J. Gore (2013). Feedback between Population and Evolutionary Dynamics Determines the Fate of Social Microbial Populations. *PLoS Biology* 11(4): e1001547.s

Sapontzis, S. (1981). A Critique of Personhood. *Ethics* 91(4): 607–18.

Schleim, S. (2012). Brains in Context in the Neurolaw Debate: The Examples of Free Will and "Dangerous" Brains. *International Journal of Law and Psychiatry* 35: 104–11.

Schultz, W. (2000). Multiple Reward Signals in the Brain. *Nature Reviews Neuroscience* 1: 199–207.

Schultz, W., P. Dayan, and P. Montague (1997). A Neural Substrate of Prediction and Reward. *Science* 275: 1593–9.

Scott, W. (1820/2008). *Ivanhoe*. Oxford: Oxford Classics.

Searle, J. (1979). Metaphor. In J. Searle, *Expression and Meaning: Studies in the Theory of Speech Acts*. Cambridge: Cambridge University Press, 76–116.

Searle, J. (2007). Putting Consciousness Back in the Brain: Reply to Bennett and Hacker, *Philosophical Foundations of Neuroscience*. In M. Bennett, D. Dennett, P. M. S. Hacker, and J. Searle, *Neuroscience & Philosophy: Brain, Mind, and Language*. New York: Columbia University Press, 97–124 (cited as Bennett et al. 2007).

Sellars, W. (1956/1991). Empiricism and the Philosophy of Mind. In W. Sellars, *Science, Perception and Reality*. Atascadero, CA: Ridgeview, 127–96. Originally printed as 'The Myth of the Given: Three Lectures on Empiricism and the

Philosophy of Mind', in H. Feigl and M. Scriven, eds., *The Foundations of Science and the Concepts of Psychology and Psychoanalysis*, Minnesota Studies in the Philosophy of Science vol. 1. Minneapolis, University of Minnesota Press, 1956.

Sellars, W. (1963/1991). Philosophy and the Scientific Image of Man. In W. Sellars, *Science, Perception and Reality*. Atascadero, CA: Ridgeview, 1–40. Reprinted from R. Colodny, ed., *Frontiers of Science and Philosophy*. Pittsburgh: Pittsburgh University Press, 1963.

Seth, A. (2014). What Behaviorism Can (and Cannot) Tell Us about Brain Imaging. *Trends in Cognitive Sciences* 18(1): 5–6.

Seymour, B., J. O'Doherty, P. Dayan, M. Koltzenburg, A. Jones, R. Dolan, K. Friston, and R. Frackowiak (2004). Temporal Difference Models Describe Higher-Order Learning in Humans. *Nature* 429: 664–7.

Shadlen, M. and W. Newsome (1996). Motion Perception: Seeing and Deciding. *Proceedings of the National Academy of Sciences USA* 93: 628–33.

Shannon, C. (1948). A Mathematical Theory of Communication. *The Bell System Technical Journal* 27: 379–423, 623–56.

Shapiro, J. (1998). Thinking about Bacterial Populations as Multicellular Organisms. *Annual Review of Microbiology* 52: 81–104.

Shapiro, J. (2007). Bacteria are Small But Not Stupid: Cognition, Natural Genetic Engineering and Socio-Bacteriology. *Studies in the History and Philosophy of Biological and Biomedical Sciences* 38: 807–19.

Shettleworth, S. (2010a). *Cognition, Evolution, and Behavior*, 2nd edn. New York: Oxford University Press.

Shettleworth, S. (2010b). Clever Animals and Killjoy Explanations in Comparative Psychology. *Trends in Cognitive Sciences* 14(11): 477–81.

Shottenkirk, D. (2009). *Nominalism and Its Aftermath: The Philosophy of Nelson Goodman*. Dordrecht: Springer.

Shulman, R. (2013). *Brain Imaging: What it Can (and Cannot) Tell Us about Consciousness*. Oxford: Oxford University Press.

Simon, H. (1962). The Architecture of Complexity. *Proceedings of the American Philosophical Society* 106(6): 467–82.

Simon, H. (1969). *The Sciences of the Artificial*. Cambridge, MA and London: MIT Press.

Simons, P. (1987). *Parts: A Study in Ontology*. New York: Oxford University Press.

Simons, P. (2000). Continuants and Occurrents, I. *Aristotelian Society Supplementary Volume* 74(1): 59–75.

Singer, P. (2009). Speciesism and Moral Status. *Metaphilosophy* 40(3–4): 567–81.

Smith, D. (2001). *Less Than Human: Why We Demean, Enslave, and Exterminate Others*. New York: St. Martin's Press.

Sober, E. (1982). Why Must Homunculi Be So Stupid? *Mind* 91: 420–2.

Sober, E. (2005). Comparative Psychology Meets Evolutionary Biology: Morgan's Canon and Cladistic Parsimony. In L. Datson and G. Mitman, eds., *Thinking With Animals: New Perspectives on Anthropomorphism*. New York: Columbia University Press, 85–99.

Sparrow, R. and L. Sparrow (2006). In the Hands of Machines? The Future of Aged Care. *Minds and Machines* 16: 141–61.

Sperber, D. and D. Wilson (1986). *Relevance: Communication and Cognition*. Oxford: Blackwell.

Sperber, D. and D. Wilson (1998). The Mapping between the Mental and the Public Lexicon. In P. Carruthers and J. Boucher, eds., *Language and Thought: Interdisciplinary Themes*. Cambridge: Cambridge University Press, 184–200.

Squire, L. (2009). The Legacy of Patient H.M. for Neuroscience. *Neuron* 61(1): 6–9.

Stern, D. (2000). *Metaphor in Context*. Cambridge, MA: MIT Press.

Steward, H. (1997). *The Ontology of Mind: Events, Processes and States*. Oxford: Oxford University Press.

Stich, S. (1983). *From Folk Psychology to Cognitive Science: The Case against Belief*. Cambridge, MA: MIT Press.

Stock, A., V. Robinson, and P. Goudreau (2000). Two-Component Signal Transduction. *Annual Review of Biochemistry* 69: 183–215.

Stout, R. (1997). Review of R. Brandom, *Making It Explicit. Mind*, New Series 108 (422): 341–5.

Strevens, M. (2012). Theoretical Terms without Analytic Truths. *Philosophical Studies* 160(1): 167–90.

Suppes, P. (1960). A Comparison of the Meaning and Uses of Models in Mathematics and the Empirical Sciences. *Synthese* 12: 287–301.

Suri, R. and W. Schultz (1999). A Neural Network Model with Dopamine-Like Reinforcement Signal that Learns a Spatial Delayed Response Task. *Neuroscience* 91(3): 871–90.

Suri, R. and W. Schultz (2001). Temporal Difference Model Reproduces Anticipatory Neural Activity. *Neural Computation* 13: 841–62.

Sutton, R. (1988). Learning to Predict by the Method of Temporal Differences. *Machine Learning* 3: 9–44.

Sutton, R. and A. Barto (1981). Toward a Modern Theory of Adaptive Networks: Expectation and Prediction. *Psychological Review* 88(2): 135–70.

Sutton, R. and A. Barto (1998). *Reinforcement Learning: An Introduction*. Cambridge, MA: MIT Press.

Tauber, A. (2013). Immunology's Theories of Cognition. *History and Philosophy of the Life Sciences* 35: 239–64.

Taylor, B. (2004). An Alternative Strategy for Adaptation in Bacterial Behavior. *Journal of Bacteriology* 186(12): 3671–3.

The Nonhuman Rights Project (2015). <http://www.nonhumanrightsproject.org>.

Tompkins, P. and C. Bird (1973). *The Secret Life of Plants: A Fascinating Account of the Physical, Emotional, and Spiritual Relations between Plants and Man.* New York: Harper & Row.

Toussaint, A., M.-J. Gama, J. Laachouch, G. Maenhaut-Michel, and A. Mhammedi-Alaoui (1994). Regulation of Bacteriophage Mu Transposition. *Genetica* 93: 27–39.

Trewavas, A. (2003). Aspects of Plant Intelligence. *Annals of Botany* 92: 1–20.

Trewavas, A. (2004). Aspects of Plant Intelligence: An Answer to Firn. *Annals of Botany* 93: 353–7.

Trewavas, A. (2007). Response to Alpi et al.: Plant Neurobiology—All Metaphors Have Value. *Trends in Plant Sciences* 12: 231–3.

Trewavas, A. (2014). *Plant Behavior and Intelligence.* Oxford: Oxford University Press.

Tribus, M. (1961). *Thermodynamics and Thermostatics: An Introduction to Energy, Information, and States of Matter, with Engineering Applications.* Princeton: D. Van Nostrand.

Turkle, S. (2006). A Nascent Robotics Culture: New Complicities for Companionship. *AAAI Technical Report Series*, July.

Unger, P. (1979). I Do Not Exist. In G. MacDonald, ed., *Perception and Identity.* Ithaca and London: Cornell University Press.

Uttal, W. (2001). *The New Phrenology: The Limits of Localizing Cognitive Processes in the Brain.* Cambridge, MA: MIT Press.

Vadasz, V. (2007). Economic Motion: An Economic Application of the Lotka-Volterra Predator–Prey Model. <https://dspace.fandm.edu/bitstream/handle/11016/4287/Vadasz.pdf?sequence=1Page>.

Van Duijn, M., F. Keijzer, and D. Franken (2006). Principles of Minimal Cognition: Casting Cognition as Sensorimotor Coordination. *Adaptive Behavior* 14(2): 157–70.

Van Fraassen, B. (1980). *The Scientific Image.* New York: Oxford University Press.

Van Gelder, T. (1995). What Might Cognition Be if Not Computation? *Journal of Philosophy* 92(7): 345–81.

Van Inwagen, P. (1990). *Material Beings.* Ithaca and London: Cornell University Press.

Velicer, G., L. Kroos, and R. Lenski (2000). Developmental Cheating in the Social Bacterium *Myxococcus xanthus*. *Nature* 404: 598–601.

Volkman, B., D. Lipson, D. Wemmer, and D. Kern (2001). Two-State Allosteric Behavior in a Single-Domain Signaling Protein. *Science* 291: 2429–33.

Von Eckardt, B. and J. Poland (2004). Mechanism and Explanation in Cognitive Neuroscience. *Philosophy of Science* 71: 972–84.

Vosniadou, S., A. Ortony, R. Reynolds, and P. Wilson (1984). Sources of Difficulty in the Young Child's Understanding of Metaphorical Language. *Child Development* 55: 1588–606.

Wagner, L. and S. Carey (2003). Individuation of Objects and Events: A Developmental Study. *Cognition* 90: 163–91.

Wang, R. and E. Spelke (2002). Human Spatial Representation: Insights from Animals. *Trends in Cognitive Sciences* 6: 376–82.

Waskan, J., I. Harmon, Z. Horne, J. Spino, and J. Clevenger (2014). Explanatory Anti- Psychologism Overturned by Lay and Scientific Case Classifications. *Synthese* 191(5): 1013–35.

Wasserman, D. (2013). Devoured by Our Own Children: The Possibility and Peril of Moral Status Enhancement. *Journal of Medical Ethics* 39(2): 78–9.

Waytz, A., K. Gray, N. Epley, and D. Wegner (2010). Causes and Consequences of Mind Perception. *Trends in Cognitive Sciences* 14: 383–8.

Webb, B. (2001). Can Robots Make Good Models of Biological Behavior? *Behavioral and Brain Sciences* 24: 1033–50.

Weisberg, M. (2007). Three Kinds of Idealization. *Journal of Philosophy* 104(12): 639–59.

Weisberg, M. (2013). *Simulation and Similarity: Using Models to Understand the World.* New York: Oxford University Press.

Wheatley, T. and J. Haidt (2005). Hypnotic Disgust Makes Moral Judgments More Severe. *Psychological Science* 16(10): 780–4.

Wilkes, K. (1975). Anthropomorphism and Analogy in Psychology. *Philosophical Quarterly* 25(99): 126–37.

Wilson, D. and R. Carston (2007). A Unitary Approach to Lexical Pragmatics. In N. Burton-Roberts, ed., *Pragmatics*. Basingstoke: Palgrave Macmillan, 230–59.

Wilson, D. and D. Sperber (2008). Relevance Theory. In L. Horn and G. Ward, eds., *The Handbook of Pragmatics*. Oxford: Blackwell, 607–32.

Wilson, M. (1982). Predicate Meets Property. *Philosophical Review* 91(2): 549–89.

Wilson, M. (1985). What Is This Thing Called "Pain"? The Philosophy of Science Behind the Current Debate. *Pacific Philosophical Quarterly* 66: 227–67.

Wilson, M. (2006). *Wandering Significance: An Essay on Conceptual Behavior.* New York: Oxford University Press.

Wimsatt, W. (1976). Reductionism, Levels of Organization, and the Mind–Body Problem. In G. Globus, G. Maxwell, and I. Savodnik, eds., *Consciousness and the Brain*. New York: Plenum Press, 205–67.

Wimsatt, W. (1987/2007). False Models as Means to Truer Theories. In W. Wimsatt, *Re-engineering Philosophy for Limited Beings: Piecewise Approximations to Reality.* Cambridge, MA: Harvard University Press, 94–132. Reprinted from M. Nitecki and A. Hoffman, eds., *Neutral Models in Biology*. London: Oxford University Press 1987, 23–55.

Wimsatt, W. (2006). Reductionism and Its Heuristics: Making Methodological Reductionism Honest. *Synthese* 151: 445–75.

Winer, G., J. Cottrell, T. Mott, M. Cohen, and J. Fournier (2001). Are Children More Accurate than Adults? Spontaneous Use of Metaphor by Children and Adults. *Journal of Psycholinguistic Research* 30(5): 485–96.

Winner, E., A. Rosensteil, and H. Gardner (1976). The Development of Metaphoric Understanding. *Developmental Psychology* 12(4): 289–97.

Wittgenstein, L. (1958). *Philosophical Investigations*, 3rd edn. G. E. M. Anscombe and G. H. von Wright, eds., trans. G. E. M. Anscombe. Oxford: Oxford University Press.

Wright, C. and W. Bechtel (2006). Mechanisms and Psychological Explanation. In P. Thagard, ed., *Handbook of the Philosophy of Science: Philosophy of Psychology and Cognitive Science*. North Holland: Elsevier, 31–79.

Wynn, K. (1996). Infants' Individuation and Enumeration of Actions. *Psychological Science* 7(3): 164–9.

Young, J. Z. (1978). *Programs of the Brain*. Oxford: Oxford University Press.

Zeki, S. (1999). Splendors and Miseries of the Brain. *Philosophical Transactions of the Royal Society: Biological Sciences* 354(1392): 2053–65.

Zink, A. and Z.-H. He (2015). Botanical Brilliance: Are Plants Decision-Makers or Elaborate Fakers? *Science* 347: 724–5.

Zwicky, A. and J. Sadock (1975). Ambiguity Tests and How to Fail Them. In J. Kimball, ed., *Syntax and Semantics Vol. 4*. New York: Academic Press, 1–36.

Index

action potential 26, 126
activity kinds 154
adaptive systems 49–54, 135–6, 153–4
    and adaptive elements 49–55
addiction 57
affect 19, 56–7, 74, 76, 87
analogical reasoning in science 15–16,
        94–5, 126–7
    see also epistemic metaphor; predicate
        extension
animal cognition 19, 23, 29, 31,
        136, 186
    and behavior
        cognitive explanations of 19, 136
        non-cognitive explanations of 19,
            23, 49, 54–5, 136–7
anthropocentrism in psychology 3–11,
        15, 23, 32, 69–71, 137, 171,
        178–80, 185
anthropomorphism 4, 167–77
    speciesism 171–80
    see also Singer, Peter
anticipation 50–9
    "anticipatal" 57
    anticipatory behavior 50–9
    see also reinforcement learning
Anti-Exceptionalism 1n1, 5–6, 61
anti-realism 32n4, 89n1, 132–3
    and nihilist metaphyics 112n9
    and relation to Literalism 89n1, 133
    see also realism and anti-realism
Aristotle 3n3, 150–1
artificial intelligence 9, 69, 156
    see also computer metaphors
ascriptions to parts
    attenuated property 101–4, 129,
        139–44, 159–60
    and brain as locus vs. subject 102–3
        see also Searle, John
    and metonymy 113
    and part–whole object hierarchies
        113–15, 145–6, 148–55, 164–5
    whole to part 91–3, 99, 101, 114, 143,
        146, 148–59, 164–5

autopoiesis 86–7
    self-organization 28

bacteria cognition 24–30, 132, 141, 170
    Brownian motion 25, 43n16
    quorum sensing 28–9
Barto, Andrew 49–59, 75, 135–6, 153
Bechtel, William 148–52, 163–4
behavior
    cooperative 28–9, 140
    cheating 27–8
    Dretske's definition of 85–6
    normatively guided 78, 138–9
behaviorism 49–50, 83–4, 98–9, 110
    in neuroscience 98–9
    see also Technical view and
        Technical-Behaviorist variant
Bennett, Maxwell 89–104, 107, 142,
        166–8
biology 15, 32, 54, 66–9, 88, 94–100,
        132, 141, 153, 184–7
    evolutionary 122n15
    micro 25
    neuro 24, 42n15
    systems 22–3
Bohr model 12n1, 126
Brandom, Robert
    deontic scorekeeping 78–9, 82
    responsiveness to reasons 78–80, 138
        see also naturalism in philosophy of
            mind
    see also space of reasons

category mistake 102
cognitive enhancement,
        pharmacological 181–2
cognitive linguistics 97n6, 124
    see also meaning vs. reference and
        meaning potentials and pointers
        to semantic space
cognitive neuroscience 91–104, 107, 156–7
complex systems 27, 140
computationalism, classical 156–65
    computer metaphors 76n10, 111, 168–9